The Social Control of Technology

The Social Control of Technology

David Collingridge

Frances Pinter (Publishers) Ltd., London
St. Martin's Press, New York

First published in Great Britain in 1980 by
Frances Pinter (Publishers) Limited
5 Dryden Street, London WC2E 9NW

ISBN 0 903804 72 7

 For information write:
St. Martin's Press, Inc., 175 Fifth Avenue, New York NY 10010
Printed in Great Britain
First published in the United States of America in 1982

ISBN 0-312-73168-X

Library of Congress Cataloging in Publication Data

Collingridge, David.
The social control of technology.

1. Technology -- Social aspects. 2. Technology
assessment. I. Title.

T14.5.C64 1980 303.4'83 80-21944
ISBN 0-312-73168-X

Reprinted in 1982

Printed in Great Britain by SRP Exeter

CONTENTS

PREFACE

This book considers one of the most pressing problems of our time – 'can we control our technology – can we get it to do what we want and can we avoid its unwelcome consequences?' The root of the manifest difficulties with which the control of technology are beset is that our technical competence vastly exceeds our understanding of the social effects which follow from its exercise. For this reason, the social consequences of a technology cannot be predicted early in the life of the technology. By the time undesirable consequences are discovered, however, the technology is often so much part of the whole economic and social fabric that its control is extremely difficult. This is the *dilemma of control*. When change is easy, the need for it cannot be foreseen; when the need for change is apparent, change has become expensive, difficult and time consuming.

The customary response to this is to search for better ways of forecasting the social impact of technologies, but efforts in this direction are wasted. It is just impossible to foresee complex interactions between a technology and society over the time span required with sufficient certainty to justify controlling the technology now, when control may be very costly and disruptive. This work proposes a new way of dealing with the dilemma of control. If a technology can be known to have unwanted social effects only when these effects are actually felt, what is needed is some way of retaining the ability to exercise control over a technology even though it may be well developed and extensively used. What we need to understand, on this view, is the origin of the notorious resistance to control which technologies achieve as they become mature. If this can be understood and countered in various ways, then the quality of our decision making about technology would be greatly improved, as the degree of control we are able to exert over it is enhanced. If a technology is found to have some unwanted social consequence, then this would not not have to be suffered for the technology could be changed easily and quickly.

The attack on trying to understand this resistance to control is a double one. A theory of decision making under ignorance – a state of deep uncertainty typical of decisions about technology – is developed which shows how decisions of this sort *ought* to be taken. Since the future is extremely uncertain, options which allow the decision maker to respond to whatever the future brings are to be favoured. Decisions, in other words, ought to be reversible, corrigible, and flexible. The theory is then illustrated in Part 1 by a series of case studies of decision making about technology in what I hope is a fruitful comparison of what ought to be done and what is actually done, to give insight into the roots of the inflexibility and resistance to control which characterizes mature technologies.

Part 2 expands various elements of the theory of decision making under ignorance to consider the role of expert advice in making decisions about technology. Such advice is essential to any technological decision and yet the traditional view of expert opinion is shown to be radically mistaken. An expert is traditionally seen as neutral, disinterested, unbiased and likely to agree with his peers. On the view proposed here, none of these qualities can be attributed. Instead, an expert is best seen as a committed advocate, matching his opinions with other experts who take a different view of the data available to them in a critical battle.

It gives me great pleasure to thank all those friends, colleagues and students who helped me in so many ways through the gestation and birth of this work: the staff, research fellows, research students and secretaries of the Technology Policy Unit, and in particular its head, Professor Ernest Braun, who first introduced me to the whole field of technology policy; and Jerome Ravetz of Leeds University and Keith Taylor of AERE Harwell for their encouragement and advice. The greatest thanks of all is to my wife Jenny, without whose enormous support in a thousand different ways – some practical, most intangible – all would still be silence.

1. THE DILEMMA OF CONTROL

There is almost universal agreement that the greatest success stories in the annals of technology are the American Manhattan Project which, in three years of research and development of an intensity never equalled, produced the world's first atomic bomb, and the American landings of men on the moon. These achievements will always stand as monuments to our ability to bend recalcitrant nature to our purposes. They reveal the depth of our understanding of the material world and our power to grasp the workings of the most complex physical systems. These successes engendered great optimism about the future benefits obtainable from technology. If the power of the atom could be understood and harnessed, and if men could walk upon the moon, what obstacles could there be to technological progress? The horizon seemed limitless – all that was required was the organization, the skill, the dedication, the tenacity, and the willingness to invest in success so characteristic of these great triumphs, and there could surely be no barrier to technology satisfying almost any human purpose. All that could stand in the way of curing disease, prolonging healthy life, feeding the hungry, providing an abundance of energy, giving wealth to the poor, was lack of will and organization.

Just this confidence in the power of technology lies behind the Green Revolution, the introduction of high yielding varieties of wheat, maize and rice in the agriculture of developing countries. As the Manhattan Project and the moon landings marked the triumph of applied physics, the development of these varieties of cereal and the fine adjustment of their characteristics to the local conditions under which they are to be grown, is perhaps the greatest achievement of applied biology. The depth of understanding of the genetics of these plants far transcends the rough and ready biological rules of thumb employed in agriculture and earlier plant breeding. The objective of the Green Revolution is to provide adequate food, and particularly protein, for the poorest sections of the population in developing countries. It has undoubtedly succeeded in increasing the total output of food and of protein, so that, for example India is now a net exporter of grains. In many parts of the world, however, the Revolution has failed in its more basic objective – although food production has increased, the diet of the poorest in the community has not improved, and sometimes has actually deteriorated. Why should this be so?

The introduction of the new high yielding varieties of the Green Revolution involves more than the substitution of one kind of seed for another. The new seeds are hybrids and so have to be purchased anew each year – the farmer no longer saves his own seed. Fertilizers, pesticides, controlled irrigation and new machinery to handle the increased crop are also required in order to benefit from the genetics of the new varieties. The Revolution has, therefore, had a profound impact on those agricultural communities which have introduced its new varieties. The benefits of the Revolution have tended to go to those with access to credit, additional land, machinery and extra labour, a group quite exclusive of the very poor. In many areas the poor have lost through the calling in of land debts by large farmers, leading to an increase in the number of landless labourers and so a drop in their wages. In addition, the increased wealth of the middle and large farmers has been accomplished by a growth in their appetite, with a corresponding increase in food prices to levels which reduce the diet of the poor. The new cereals have also tended to replace pulses, which have traditionally been important in giving a balanced protein intake to the largely vegetarian poor. Attempts by governments to exploit the new varieties in making their countries self-sufficient in food have also worsened the lot of the poorest by raising prices, and further reducing pulse production.

In a review of the consequences of the Green Revolution, Pauline Marstrand states the problem concisely:

> Increased productivity of food crops must include development of better varieties which respond to improved husbandry, but raised productivity will not follow automatically the introduction of new varieties. Unless governments deliberately counteract the tendency of *any* new technology to increase inequalities between groups in society, by controlling access to credit, prices of produce, accumulation of land and by discriminating in favour of the poorest people, such introductions may increase food production, but will not prevent malnutrition or even famine.[1]

What can be learned from a comparison of the great success of the Manhattan Project and the moon landings and the failure of the Green Revolution to meet its objectives of feeding the poor? As mentioned before, the development of the new crop varieties cannot be counted less of an accomplishment than the development of the atomic bomb or the landing of men on the moon. The atomic bomb exploded; men landed on the moon, walked around and returned to Earth; and the new cereals produce heavier yields than earlier varieties. The difference between success and failure in no way stems from a technical miscalculation on the part of the technologists. The key difference here is that the objectives given to the successful programmes were purely technical. Success for the Manhattan Project was a bomb which exploded with more than a particular force.[2] Success for the moon programme was the landing of a piece of hardware carrying a man and its safe return to the Earth. The Green Revolution is quite different. Its objective was not a technical one, but a human one. The Revolution's aim was not the breeding of high yielding cereals, but the bringing of food to the very poor.

The Green Revolution is highly successful in the sense of its technical achievements, but a total failure because these vast efforts have all been in quite the wrong direction. As Marstrand observes, providing more food without changes in the social conditions which determine how the benefits from this are divided cannot bring food to the poorest. This is what has robbed the Revolution's efforts of success; the lack of understanding of how the technical products of the Revolution interact with the society using them. The lesson from this is salutary: our understanding of the physical and biological world in which we live is extremely deep, and provides us with the means for the production of all kinds of technical marvels; but our appreciation of how these marvels affect society is parlous. Ask technologists to build gadgets which explode with enormous power or to get men to the moon, and success can be expected, given sufficient resources, enthusiasm and organization. But ask them to get food for the poor; to develop transport systems for the journeys which people want; to provide machines which will work efficiently without alienating the men who work them; to provide security from war, liberation from mental stress, or anything else where the technological hardware can fulfil its function only through interaction with people and their societies, and success is far from guaranteed.

It is this huge disparity between our technical competence and our understanding of how the fruits of this competence affect human society which has given rise to the widespread hostility to technology. The critics have no shortage of examples. Modern medicine and hygiene has reduced the death rate in developing countries, but doing so has generated an uncontrollable increase in population. Modern technology has conferred on the inhabitants of developed countries an unprecedented standard of living, but only at the cost of environmental damage and resource consumption on such a scale as to make society non-sustainable. Technological progress enables workers to produce far more than previously, to the benefit of all, but only because the work has been robbed of all individuality and reduced to meaningless repitition. The technology of nuclear weapons may have prevented war for a few decades, but at the cost of vastly greater destruction when the war finally comes. Food production has increased through the use of chemicals, but at the cost of the future collapse of agriculture due to damage to the soil and to its supporting ecosystem. Modern transport is so cheap that many can enjoy their leisure in beautiful parts of the world, but at the cost of reducing the local population to serfdom and forcing their surroundings into the deadly uniformity of hotel fronts.

Thus, technology often performs in the way originally intended, but also proves to have unanticipated social consequences which are not welcome. It is this problem to which the present work is devoted. What I wish to ask is how can we make decisions about technology more effectively; how can we get the technology we want without also having to bear the costs of such unexpected social consequences, and how can we avoid technologies which we do not want to have? To put it in another way: the problem is how technology can be controlled in a better way than at present. There is a central problem concerning the control of technology which we may now focus on, which I shall refer to as the *dilemma of control.*

Two things are necessary for the avoidance of the harmful social consequences of technology; it must be known that a technology has, or will have, harmful effects, and it must be possible to change the technology in some way to avoid the effects. In the early days of a technology's development it is usually very easy to change the technology. Its rate of development and diffusion can be reduced, or stimulated, it can be hedged around with all kinds of control, and it may be possible to ban the technology altogether. But such is the poverty of our understanding of the interaction of technology and society that the social consequences of the fully developed technology cannot be predicted during its infancy, at least not with sufficient confidence to justify the imposition of disruptive controls. The British Royal Commission on the Motor Car of 1908 saw the most serious problem of this infant technology to be the dust thrown up from untarred roads. With hindsight we smile, but only with hindsight.

Dust was a recognized problem at the time, and so one which could be tackled. The much more serious social consequences of the motor car with which we are now all too familiar could not then have been predicted with any certainty. Controls were soon placed on the problem of dust, but controls to avoid the later unwanted social consequences were impossible because these consequences could not be foreseen with sufficient confidence.

Our position as regards the new technology of microelectronics mirrors that of the Royal Commissioners in 1908. This technology is in its infancy, and it is now possible to place all kinds of controls and restrictions on its development, even to the point of deciding to do without it altogether. But this is a freedom which we cannot exploit because the social effects of the fully developed technology cannot be predicted with enough confidence to justify applying controls now. Concern has been expressed about the unemployment which may result from the uncontrolled development and diffusion of microelectronics, but our understanding of this effect is extremely limited. The future development of the technology cannot be foreseen in any detail, nor can its rate of diffusion. Even if these were known it would be impossible to predict the number of workers displaced by the new technology in various sectors. This depends upon a whole bundle of unknown factors; the demand for the displaced labour from other sectors of the economy, the number of jobs created by the economic savings from the new technology, the number of jobs involved in making, developing and servicing microelectronics and so on. It is hardly surprising to find that forecasts of the effects on employment of the uncontrolled development of microelectronics cover a huge range.[3]

It is clear that our understanding of how this infant technology will effect employment when it is fully developed is so scanty that it cannot justify the imposition of controls on its development and diffusion. It will not do to suggest playing safe and imposing such controls to avoid the unemployment which *may* result from the new technology. This is to forego all the benefits from microelectronics for the avoidance of the unquantifiable possibility that it will cause serious unemployment. Such caution would effectively eliminate technological change.

The motor car in 1908 and microelectronics now are typical of technologies in their infancy. They may be controlled easily in all sorts of ways, but control can only be arbitrary. Our understanding of the interactions between technology and society is so poor that the harmful social consequences of the fully developed technology cannot be predicted with sufficient confidence to justify the imposition of controls. This is the first horn of the dilemma of control.

The second horn is that by the time a technology is sufficiently well developed and diffused for its unwanted social consequences to become apparent, it is no longer easily controlled. Control may still be possible, but it has become very difficult, expensive and slow. What happens is that society and the rest of its technology gradually adjust to the new technology, so that when it is fully developed any major change in the new technology requires changes in many other technologies and social and economic institutions, making its control very disruptive and expensive.

The motor car may again be taken to illustrate the point. As this technology has gradually diffused adjustments have been made by other modes of transport. The provision of buses and passenger trains has adjusted to the existence of more and more private motor cars, and villages have been urbanized and outer suburbs grown as travel has become cheaper and more convenient. The ability of more and more workers to move around easily has led to more and more offices and factories moving from city centres. The importance of crude oil for the production of petrol has also grown, and the chemical industry has adjusted to the existence of the motor car by learning how to exploit the residues from crude oil after petrol extraction. The existence of the motor car also made alternatives to the internal combustion engine very unattractive so that, for example, very little research and development on electric motors for road transport has been done. The list could be extended indefinitely.

Imagine now that severe controls must be placed on the use of motor cars because, for example, crude oil has suddenly become very expensive and scarce, or because some pollutant of motor car exhaust is suddenly found to be many times more toxic than previously thought. Society has adjusted to cheap transport in its siting of houses, factories and offices, so sudden restrictions on transport would inevitably be very expensive. Large parts of the economy would virtually collapse. A significant shift from private cars to buses and trains would be impossible in the short term because their number has adjusted to the existence of a transport system dominated by the private car. It would take many years, and huge investment before buses and trains could carry anything but a tiny fraction of those previously using motor cars. Confident of crude oil supplies, the chemical industry has invested very little research and development into using any other feedstock than crude oil residues, and so many years would be needed before it could adjust to the new situation, making the change extremely expensive. Similarly, little work has been done on substitutes for the petrol-fuelled internal combustion engine, so that a long time would be needed before substitutes could be used on a large scale.

The example is an extreme one, but as such it makes the point: the interaction between technology and society works in such a way as to make the severe control of a major established technology very expensive and necessarily slow. As for the motor car, so for microelectronic technology. By the time this is fully developed and diffused, all sorts of adjustments will have been made which will make its control very costly and slow.

The dilemma of control may now be summarized: attempting to control a technology is difficult, and not rarely impossible, because during its early stages, when it can be controlled, not enough can be known about its harmful social consequences to warrant controlling its development; but by the time these consequences are apparent, control has become costly and slow.

The concern of this book is the efficient control of technology and so its natural starting place is this dilemma of control. The customary response to the dilemma concentrates on the first horn, and amounts to the search for ways of forecasting a technology's social effects before the technology is highly developed and diffused, so that controls can be placed on it to avoid consequences not wanted by the decision makers and to enhance the consequences they desire. On this view, the key problem about the control of microelectronics is the lack of a powerful forecasting device able to give firm information about the employment consequences of this technology's uncontrolled development. Many of the forecasting techniques which have been devised are extremely dubious, but what lies behind the expenditure of effort in this direction is a serious misconception about the quality required of forecasts. To be of any use, a forecast of future unwanted social effects from a technology now in its infancy must command sufficient confidence to justify the imposition of controls now. It is not enough for the forecast merely to warn us to look for bad social consequences of a particular kind in the future, because by the time they are discovered their control may have become very difficult and expensive. The prediction of social effects with such confidence demands a vastly greater appreciation of the interplay between society and technology than is presently possessed. I doubt that our understanding will ever reach such a sophisticated level, but even if this is possible it will only be as the outcome of many years of research. Thus even an optimistic view leaves us with the problem of how to improve the control of technology in the period needed for the development and testing of adequate forecasting methods. A pessimist like myself regards this period as effectively infinite.

The only hope seems to be in tackling the other horn of the dilemma of control. If the harmful effects of a technology can be identified only after it has been developed and has diffused, then ways must be found of ensuring that harm is detected as early as possible, and that the technology remains controllable despite its development and diffusion. On this view, the key problem of controlling microelectronics has nothing to do with forecasting, but is to find ways of coping with any future unemployment which it may generate, and ways of avoiding overdependence on the technology which would make it difficult to control.

This is an enterprise requiring nothing less than a wholly new way of looking at decisions concerning technology. In attempting to develop this new viewpoint, many starting places suggest themselves. Insight might be obtained from control theory, but this proves disappointing as it is exclusively concerned with control problems which are vastly better organized, structured and understood than the problems which arise in attempting to control technology. The same may be said of systems theory and of conventional economics. What might be called the political theory of technological decision making is doubly flawed as a starting point, for not only is it very poorly developed, but it asks the wrong question. The theory attempts to understand *why* decisions about technology were made in the way that they were; but our problem is about how such decisions *ought* to be taken if effective control is to be exercised over technology. This last comment suggests that a more promising starting point may be found in decision theory which considers just this question – how decisions ought to be taken. As we shall see in the next Chapter, decision theory needs to be extended in a quite dramatic way before it can provide assistance in the quest for ways of coping with the second horn of the dilemma of control, but at least it provides a starting point for the search.

In writing this book I have been aware of the tension between the development of a theory of decision making able to cope with the kind of decision problems which arise in the control of technology, and the practical application of this theory. An equilibrium has eventually been reached, for good or ill, which gives prominence to the practical application of theory and where purely theoretical discussion has been reduced to the very minimum necessary to appreciate these applications. Chapter 2 develops a theory of decision making which calls for decisions taken under the circumstances typical of decisions about technology to be reversible or flexible, so that if they should be discovered to have been wrong, something can be done to remedy this.

The essence of controlling technology is not in forecasting its social consequences, but in retaining the ability to change a technology, even when it is fully developed and diffused, so that any unwanted social consequences it may prove to have can be eliminated or ameliorated. It is, therefore, of the greatest importance to learn what obstacles exist to the maintenance of this freedom to control technology. This is the task of Part 1 where the theory of Chapter 2 is applied to a number of historical case studies.

Theory then re-appears in Chapter 10 which opens Part 2 and which considers the relationships between fact and value which are of importance to the theory developed in Chapter 2. Some theoretical innovation is called for to justify what has been assumed throughout Part 1, that decisions can be revealed to be mistaken by the discovery of facts. This part of the theory is then applied, again through case study material, to questions concerning the role of experts in decision making about technology, the nature of disagreement between experts and the need for public debate of major issues in the adoption and deployment of technology.

My reason for minimizing the discussion which is purely theoretical is a double one. The theory has been presented elsewhere,[4] albeit in an unfinished way, and I hope to provide a more polished version shortly; and in the present work I hope to show that the new theory has sufficient practical promise to deserve attention.

References

1 Marstrand (1979). For fuller accounts see: Berg (1973), Dasgupta (1977), Griffin (1974) and Poleman and Freebairn (1973).

2 The Project will be discussed more fully in Chapter 8. Notice that its objective was not to win the War, in which case it would have to be counted a failure as the bomb came too late to have any serious effect on the War.

3 Compare, for example, Sleigh *et al.* (1979) with International Metalworkers Federation (1979).

4 Collingridge (1979).

Bibliography

A. Berg (1973), *The Nutrition Factor*, Brookings Institute.

D. Collingridge (1979), *The Fallibilist Theory of Value and Its Application to Decision Making*, Ph.D. Thesis, University of Aston.

B. Dasgupta (1977), 'India's Green Revolution', *Economic and Political Weekly, 12,* 6, 7, 8, February 240-59.

K. Griffin (1974), *The Political Economy of Agrarian Change: An Essay on the Green Revolution,* Macmillan.

International Metalworkers Federation (1979), *Effects of Modern Technology on Workers*, IMF, Geneva.

P. Marstrand (1979), 'Shadows on the 70s: Indicative World Plan, Protein Gap and the Green Revolution', in T. Whiston (ed.), *The Uses and Abuses of Forecasting,* Macmillan.

T. Poleman and D. Freebairn (eds.), *Food, Population and Employment: The Impact of the Green Revolution*, Praeger.

J. Sleigh *et al* - (1979), *The Manpower Implications of Microelectronic Technology*, HMSO.

2. DECISION MAKING UNDER IGNORANCE

In the previous Chapter it was argued that the control exercised over technology can be enhanced by tackling the second horn of the dilemma of control, but not the first. The central research problem concerning the control of technology is not to find better ways of forecasting the social effects of technology; it is to understand why it is that as technologies develop and become diffused, they become ever more resistant to controls which seek to alleviate their unwanted social consequences. When this effect is understood, we may learn how to guide technological developments so as to avoid, or at least reduce, the build up of this resistance to control. Our concern is, therefore, the normative one of how decisions of this kind *ought* to be taken. How some of these decisions are actually taken will also be of interest later on, but only as examples with which to explore normative ideas. The normative nature of our inquiry suggests that a theoretical starting point might be provided by decision theory which considers how decisions of various kinds ought to be made.

Decisions which belong to the real world, and not to the imaginations of textbook authors, present a bewildering richness of variety which must be organized in some way or other. To make any kind of progress, one kind of decision problem needs to be separated out as especially simple, and decision theory asked to provide a solution to this class of transparent problems. Once decision theory can show how decisions of this peculiarly simple, and even artificial kind ought to be made, it can then develop, devouring decision problems of ever increasing complexity and ever increasing closeness to real life problems.

This, of course, is what has happened. The kind of decision problem identified as particularly simple has the following features: there is a single decision maker; the decision maker has a single objective; the decision maker knows all the options which are open to him; there is an objective function whose values offer a measure of the extent to which the decision maker's objective is reached; and the outcome of each option is known and determines a value of the objective function. This is the structure of the decision problems picked out as simple, and the solution to such problems provided by decision theory is equally straightforward. The decision maker seeks an objective and the objective function measures the level to which the objective is satisfied, so he ought to select that option which yields the optimum value of the objective function. If the objective and objective function are simply related, this amounts to maximizing the function: if inversely related, to minimizing it. Thus is born the central concept of decision theory as it has developed to date, the idea of the rational decision maker as one who seeks to optimize what he receives as a consequence of his decisions. It is easy to understand the grip which this chosen starting point exercises on the mind of decision theorists. After all, what is more natural than that a person seeks to get as much of what he wants as he can?

Having selected a simple variety of decision problem and provided a solution to it, decision theory has gone on to tackle more complex problems. The simple variety of decision problem described above is said to be one under *certainty* because enough is known about the outcome of each of the decision maker's options to determine a value for the objective function for each option. There are cases, however, where the outcome of at least one option cannot be identified, but two or more possible outcomes exist. If an objective probability distribution over the outcomes is known, then the decision is said to be one under *risk*, and decision theory offers the so-called Bayesian rule that the decision maker ought to optimize the *expected* value of the objective function. In a yet more complex case, a decision is said to be under *uncertainty* if there is inadequate information to determine an objective probability distribution. Here the customary solution offered by decision theory is the use of subjective probabilities, with the same Bayesian rule as before.[1]

These developments enable decision theory of this kind to which we may extend the label *Bayesian*, to cope with decision problems where factual information is more and more sparse, but we are still dealing with a single decision maker with one objective. Multiattribute decision theory attempts to tackle decisions involving a number of objectives.[2] Where a group of decision makers are involved, game theory, as developed by von Neumann and Morgenstern, attempts to tackle decision problems where members of the group are rivals.[3] Where group decisions of a political kind are involved, Rawls, Harsanyi and others have tried to show how decisions ought to be made given the Bayesian assumption that each individual seeks to optimize his own position, a problem also tackled in a slightly different way by welfare economics.[4]

The theoretical developments from the simple starting place are very impressive. Yet it is my contention that Bayesian decision theory can never be developed to give guidance about the sort of decisions typical of those which arise in controlling technology. The one argument I wish to urge here on this point is that in controlling technology there is typically inadequate factual information available for the application of any Bayesian rules.[5] Existing terminology is, perhaps deliberately, very confusing, with the term 'uncertainty' being used to cover all cases where factual information falls short of making the decision one under risk. The Bayesian theory of decision making under uncertainty, however, only applies to decision problems where all of the possible outcomes, or states of nature, relevant to the decision are identifiable – where what is missing is a complete probability distribution over these states of nature. This leaves the whole class of decision problems where not all the relevant states of nature can be identified. These are also covered by the description 'uncertainty', although no Bayesian decision procedure can be applied to them.

To mark the distinction between cases where some kind of Bayesian optimization is and is not possible, I propose to confine the term 'uncertainty' to the former, and to employ the term 'ignorance' to the latter. The traditional spectrum of decisions under certainty : risk : uncertainty is therefore expanded to certainty : risk : uncertainty (in the restricted sense) : ignorance. What is needed to show that this is more than a purely academic exercise is a class of important decisions which have to be taken under ignorance. This is provided by the decisions which typically have to be made about controlling technology. This is best seen by considering a couple of cases.

Control of environmental lead

Lead is a well known poison of the nervous system and yet a typical British city dweller is in contact with the metal through the air he breathes, where it originates largely from motor-car exhausts; from the food he eats, most of which is the result of contamination during processing, e.g. in canning; and in the water he drinks, from lead piping. Lead is excreted from the body fairly rapidly so the metal does not accumulate indefinitely, but reaches an equilibrium concentration throughout the body which is far below the levels required to produce the symptoms of classical lead poisoning.

Nevertheless, recent experimental and epidemiological work has pointed to the possibility that even these low levels may have subtle health effects, particularly on the central nervous system. The group particularly at risk here are pre-school children whose nervous systems are developing rapidly and are sensitive to any toxin. It is possible that a proportion of children of this age with quite normal exposure to lead in the environment have impaired central nervous function which reveals itself in lower IQ scores, difficulty in learning and behaviour problems. The evidence on this point can hardly be reviewed here, but suffice it to say that, despite a considerable research effort, there is at the moment no consensus in the scientific community about the existence of these effects.

Exposure to lead may be lowered in a number of ways; by reducing the amount of lead added to petrol as an anti-knock agent; reducing lead in food by, for example, replacing lead solder in canning; and reducing the levels of the metal found in domestic water by water treatment or the removal of lead piping. None of these options is, however, cheap. The decision must be made, in the light of the scientific evidence mentioned above and the various control costs, whether to lower environmental lead levels by invoking one of these controls now, or to delay the decision until the scientific community eventually reaches a consensus about the health effects of lead. This is a very common decision problem which yields to Bayesian optimization only if a number of conditions can be met. The problem can be represented in the customary form of a decision tree shown in Figure 2.1. For a Bayesian solution to be possible the following must be known:

(a) the reduction in damage to health per year which would be achieved by the control, measured in the same units as the cost of the control, which must also be known;
(b) the time t needed for a consensus to be reached about the health effects of lead in the environment and;
(c) the probability that lead will be found to be harmful by time t.

If these are known, it is easy to identify which course of action – 'apply control now' or 'delay decision' yields the greatest expected pay-off.

None of these vital factors is known. If lead proves to be a hazard to health, it will still not be possible to measure the improvement in health brought about by the control in money terms so that it is comparable to the cost of control; there is simply no method by which this can be done. Secondly, there is no way of knowing how long a scientific consensus will take to emerge, and no way of rationally assigning a probability to the consensus being that lead is or is not a health hazard. The Bayesian approach, therefore fails in this case, which provides us with an example of a decision which must be taken under ignorance.

Figure 2.1

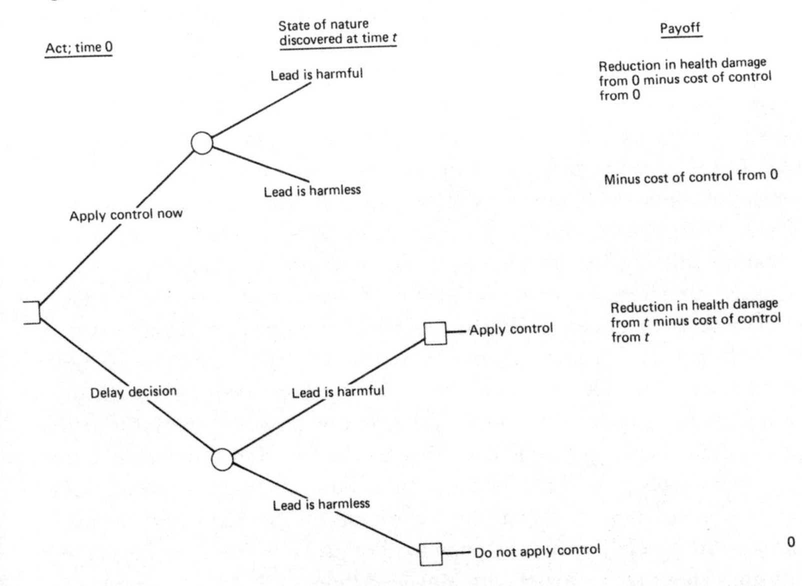

Before leaving this example, let it be noted that there is nothing peculiar about it, rather it is typical of decisions about the imposition of controls on environmental hazards, and indeed, typical of decisions dependent on research findings. Research is generally slow at reaching a consensus. Occasionally one remarkable breakthrough will settle the matter at issue once and for all, but a more general picture is that of a slowly emerging consensus in the community of experts that the growing body of data can only be interpreted in one way. In such a case the decision maker inevitably faces the same sort of problem about the control of lead. Should he act before a final consensus is reached, or should he delay his decision until it can be settled on the basis of the consensus? For the same reasons as before, such a decision will be one under ignorance.

Energy supply and economic growth

Energy is a crucial factor in economic activity and many fears have recently been expressed that energy supplies may one day restrict economic growth very severely. A great deal of attention is, therefore, now given to energy planning. In much of this work there is an underlying objective, sometimes explicit sometimes not. What is sought is the cheapest energy system which will not limit economic growth, with constraints on this objective to ensure that such things as the environment and international relations are protected.[6] Having this objective function does not, however, mean that the energy system can be optimized. As before, the problem arises from lack of knowledge, not scientific knowledge this time so much as knowledge of the future. Because of the enormously long lead times of energy technologies, choices made now about what technologies to explore and foster today will determine which of these technologies will be available for the energy system of 30–40 years hence. But it is quite impossible to forecast the energy situation this far into the future to see what energy technologies will be required to meet the desired objective. Once again, then, we are faced with a decision under ignorance. To select the optimum energy system through time we would require knowledge of the future 30–40 years so that we could identify now the technologies needed then. Such forecasting is not possible, so optimization is not possible; the information required is simply not available. Hence, the decision problem is under ignorance. Later we shall discuss some of the attempts made to tackle this problem.[7] For now, let the nature of the problem be recognized and also its generality, for the same kind of problem faces anyone making decisions about technologies with long lead times.

These two examples show the existence of a class of extremely important decisions which have to be made under ignorance, and so cannot be treated by Bayesian theory. What is required therefore is some way of extending decision theory to cover decisions under ignorance. Once this is developed a theoretical framework will exist for the examination of decisions about the control of technology. This is a task for the remainder of this Chapter.

The theory originates from the realization that decisions under ignorance cannot be justified. A compelling feature of Bayesian decision theory is that it shows how some decisions may be justified. If an agent freely adopts the objective X, and if the objective function Y measures the extent to which this objective is fulfilled, then the agent is justified in choosing so as to optimize the value of Y. For a decision which has to be taken under ignorance, however, there is simply not sufficient factual information available for such optimization. At first, this may seem to imply a depressing scepticism, until it is remembered that there are two very distinct traditions in viewing rationality and rational behaviour. The historically dominant tradition, and one to which Bayesian decision theory firmly belongs, sees rationality as the acceptance only of opinions which are justified, and sees a rational agent as one who justifies what he does.

The rival fallibilist tradition denies the possibility of justification, and sees rationality as the search for error and the willingness to respond to its discovery.[8] It does not follow, therefore, that because decisions under ignorance cannot be justified, they cannot be made in a rational way. What is needed is a fallibilist account of rationality for these decisions.

Since a decision under ignorance cannot be justified, it is always possible to find that it is in error. What is meant by this is that the decision maker may always discover new information which, given his objectives, shows that some other option would have served him better than the one he has chosen. If this information comes to light, I shall speak of the original decision as being mistaken, wrong or in error, notwithstanding that this error could not have been foreseen or avoided at the time the decision was made. In this way to call a decision mistaken is not a reflection upon the decision maker who may have done his best with an intractable decision problem.

A full account of when a decision must be regarded as falsified by a fact awaits a theoretical discussion which is best left to Part 2. In the absence of this theoretical account, however, clear cut cases of the factual falsification of decisions may be pointed to. In the real world the discovery of mistakes is an all too frequent event. Many decisions are made in response to some problem, pressing at the moment or expected to press, and are shown to be wrong when the problem either disappears or fails to materialize. Hitler's famous decision to keep the Panzers from the D-day beaches was made because he expected a more serious landing in the Calais area. This never happened, so we think of his decision as mistaken, even though we might refrain from blaming him for poor decision making. Other decisions may be discovered erroneous when they have consequences regarded by the decision maker as bad which were never taken into account. The extensive use of chemical pesticides leading to resistant pests, attack by new pests because of the unbalancing of the ecosystem, and biological magnification of residues might be an example here. A third class of decision is discovered to be wrong when the objective is missed by a wide margin. A decision by a company seeking to maximize profit which leads to a loss, or even to collapse of the company, is shown, by these consequences, to have been mistaken.

Under conditions of ignorance it will always be possible to engage in special pleading about what appears to have been a wrong decision. Given all the uncertainties of war, perhaps Hitler's witholding of the Panzers was the best move; perhaps the early use of chemical pesticides was correct, after all, because of the overall benefits which accrued; perhaps the company's employees will really enjoy their enforced leisure, and so on. But if error is to be recognized and guarded against in the decisions we make, such special pleading must be recognized as what it is; the age old attempt to defend vested interest and reputation against criticism. This will become clearer after Chapter 8 which considers problems of monitoring decisions and ways of evading the results of monitoring.

Any decision made under ignorance may, to recapitulate, be discovered to have been in error. This may be depressing to those with over high expectations of decision making, but it provides the key to the solution of how decisions under ignorance should be tackled. If the possibility of error is unavoidable, we must learn to live with it. We must be prepared to discover our mistakes and to remedy them. In making a decision under ignorance two things are essential; the ability to discover information which would show the decision to be wrong, and the ability to react to this information if it ever comes to light. Whatever decision is made, factual information which would reveal it to be wrong must be searched for, and if it is found, the decision must be revised. Revision is a problem, because any decision involves the investment of resources, not all of which can generally be recovered when the decision is changed. Decisions which, if mistaken, can be discovered as such quickly, and which can be revised with the rescue of a large fraction of their invested resources should be favoured when operating under ignorance. In choosing such options, the decision maker is ensuring that his long-term payoff is insensitive to error. This makes sound sense. Where there is no way of guaranteeing freedom from error, decisions whose payoff is insensitive to error should be favoured.

An immediate consequence of this view is that a decision under ignorance cannot be seen as a point event; it must be seen as a process. On the Bayesian approach decisions are point events, so that given the state of knowledge about outcomes, options and payoffs at some particular time and the decision maker's preferences at that time, his decision can be assessed as rational or irrational. The process by which the information was acquired and the preferences developed is of no interest; once they are settled a Bayesian algorithm can be applied to give the correct decision. It is quite otherwise for decisions under ignorance. Given the information and objectives of one time it is not possible to determine the correct decision, nor to assess the real decision as rational or irrational. This is not a function of the point decision, but depends partly on how the decision maker is prepared to act in the future. Will he, that is, search for information able to show that his original decision is wrong, and if he finds it, will he be able to revise his decision in the light of the discovery? The decision is not just the selection of an option, but this plus a search for information in the future and the ability and resolve to act on this information if it is found. The decision is, in short, a process and not a point. This seems a much more realistic view of complex decision making than that offered by Bayesian theory. It also promises power for the theory of decision making under ignorance, because the more aspects of real decision making which can be explained theoretically, the stronger the theory. Bayesian theory is weak because it says nothing about what comes before or after the point decision.[9]

What follows is a framework for making decisions under ignorance and analysing historical decisions of the same kind. The central idea is that under conditions of ignorance a premium ought to be placed on decisions which can swiftly and easily be recognized as wrong, and which are easy to correct. In general, an option with these characteristics will be more expensive than an option for which the discovery and correction of error is difficult, so that at the heart of many decisions under ignorance is a trade-off between the ease with which mistakes can be detected and eliminated, and cost. This is not to suggest that an optimal trade-off can be identified, as when Bayesian theory can be applied. In cases of ignorance there is simply not enough information to identify this optimum. It is impossible to calculate the benefit or expected benefits from the option chosen, so this cannot be compared with its cost. What can be done, however, is to compare the additional cost of the option with the additional ease of its correction. An example might be useful here.

Consider the choice between various motor cars with different instrumentation, controls, safety features, and price. The problem is to decide whether additional features are worth the extra expense. On the Bayesian model this requires detailed knowledge of the conditions which the car will meet during its lifetime. The roll bar, for example, may be worth the extra cost if the car is one day going to turn over, but not otherwise. Not only is it impossible to predict such events, it is impossible to assign a probability in any rational way, so the Bayesian approach to the decision problem is, again, too demanding of information to be of any help.

In driving a car, errors must be recognized and action taken to avoid them. Thus we can say that if a car has a speedometer, driving mistakes are easier to detect than in the same car without the instrument. The driver has more information, and this may sometimes reveal errors in the driving decisions he has made. If a car has more powerful brakes than one similar in other respects, it is obviously easier to correct mistakes when driving the former. Not all driving errors require braking, but some may, and so extra braking power makes mistakes easier to eliminate. Similarly a roll bar makes some errors less expensive when they are unavoidable. In this way it is possible to identify three categories of equipment which enhance the detection and elimination of errors; information devices, control devices, and devices for reducing error costs. This is possible even when the use to which these various devices will be put throughout the car's history is unknowable. The effectiveness of these items must then be compared with their cost.

In what follows I wish to develop four equivalent ways of looking at decisions under ignorance. Such decisions should be highly corrigible; should involve the choice of systems which are easy to control; should keep future options open; and should be insensitive to error.

1. The Corrigibility of Decisions

A decision is easy to correct, or highly corrigible, when, if it is mistaken, the mistake can be discovered quickly and cheaply and when the mistake imposes only small costs which can be eliminated quickly and at little expense. The essence of decision making under ignorance is to place a premium on highly corrigible options. The various elements which are relevant to an option's corrigibility are as follows.

(a) Monitoring

In decision making under ignorance it may happen that the chosen option has unexpectedly bad consequences, which may be so bad as to render the chosen option inferior to a rival option, given the decision maker's objective. An essential feature of such decision making is the search for such consequences; the search for error in the original decision. The scrutiny of a decision's real consequences with the aim of finding error may be termed *monitoring* the decision. The minimum period from the decision to the discovery of error may be called the *monitor's response time*. Options which can be monitored cheaply should be favoured because monitoring costs are inescapable in decision making under ignorance. In the rest of the discussion of this Chapter monitoring costs will, however, be reckoned as zero, which is a simplification which can often be made in real world decision problems. The alterations necessary to accommodate non-zero monitoring costs should, however, be fairly straightforward. Options which have a low monitor's response time should also be favoured. If a decision is mistaken, it pays to discover the mistake quickly. The proper function of forecasting is here. Any forecast used in making a decision under ignorance may be wrong, so it is essential to monitor a decision based on such a forecast, and to retain the ability to alter the decision if error is discovered. Without such safeguards, forecasting can lead to disaster. Forecasts tell the decision maker what he should look out for in monitoring his decision.

(b) The cost of error

When a wrong decision has been made costs, though not necessarily monetary ones, are imposed; indeed it is these costs which constitute the error. The cost of an error is the decision's *error cost*, which is generally a function of time. If no corrective action is taken, we may speak of *uncontrolled error cost*, remedial action reducing this to a *controlled error cost*. Options with a low controlled error cost should be favoured if a decision has to be made in a state of ignorance. If the chosen option proves to be mistaken the mistake need not involve the bearing of great costs.

(c) Time for correction

A remedy for a discovered error generally takes time to operate fully. The period from the remedial action to the elimination of error cost may be termed the *corrective response time*. The sum of this and the monitor's response time is the *gross response time*. Options with a low corrective response time should be favoured in making decisions under ignorance. There are three reasons for this. Remedying a mistaken decision quickly means that error costs are eliminated quickly and benefits from an improved decision are obtained early; remembering that early benefits are to be valued more highly than postponed benefits. Secondly, ignorance often extends to the effectiveness of the imposed remedy. When this is so, a remedy with a low corrective response time may be discovered to be ineffective more quickly than a remedy with a high corrective response time. This means that another, and perhaps more effective remedy can be substituted more quickly. The sooner the substitution, the lower the error cost. A low corrective response time thirdly leaves the decision maker with more options once the discovered mistake has been remedied, as will be shown below.

(d) The cost of remedy

The cost of applying a remedy for a mistaken decision may be called the *control cost*. Low control cost is obviously a desirable feature of decisions made under ignorance, but, in addition, where the effectiveness of the remedy is not known favour should be given to remedies having a high variable : fixed cost ratio. If the remedy is found to be ineffective, then all the fixed costs are generally lost, but some of the variable costs will be avoidable once another remedy is substituted.

Having identified the facts relevant to the corrigibility of a decision made under ignorance, the relationship between them may now be considered. It is useful to regard the first three factors as determining the decision's corrigibility, control cost then being seen as the cost of imposing a remedy which exploits the decision's corrigibility. A number of remedies will normally exist once monitoring has revealed a mistake, but often their gross response time and controlled error costs vary together. When this happens these two factors pull in the same direction and the set of remedies may be said to be *well behaved*. Ordinal measures for the decision's corrigibility can then be applied; the two simplest being $[\mathrm{MIN}\ (\text{response time})]^{-1}$ and $[\mathrm{MIN}\ (\text{controlled error cost})]^{-1}$.

The idea behind these measures is quite simple. The corrigibility of an option measures how easily it *may* be corrected, paying no regard to the cost of correction. The two factors which determine corrigibility are gross response time and controlled error cost. When these vary together for a set of remedies, a low controlled error cost being associated with a low gross response time, the ease of control may be measured by the least time in which it is possible to remedy a discovered error, or the lowest error cost which the mistake can impose. These measures are ordinal because they enable decisions to be ordered with respect to corrigibility, although differences and ratios of corrigibilities so measured have no significance. We can say that one option is more corrigible than another, but not by how much.

A final point which should be stressed once more, is that corrigibility generally costs money, so that there is a tension between ease of correction and control costs. This tension is at the heart of many decisions made under ignorance.

An example might now be valuable, and no harm will be done by considering a problem which will be encountered again in more detail in Chapter 7. Consider a firm whose product expects a growing market, but one which cannot be forecast with sufficient accuracy to enable an optimum development of production plant to be planned. Decisions about future levels of output and capacity are, therefore, ones under ignorance. The firm's objective is to gain as large a market share as possible. The decision to be made is whether to organize production in large units or small ones, producing a quarter of the output of the large. It takes one year to build and commission a small unit, three years for a large unit, and contractual arrangements with plant suppliers mean that, for some years at least, all plants must be of the same type. We may assume that there are economies of scale, unit costs being less for large than for small units.

The objective of the firm is to maximize its market share (though, of course, this may be subsumed under a more general objective such as profit maximization), and decisions about the number of production units and their size place restrictions on output and so on market share. A decision about production capacity is, therefore, shown to be wrong if it is discovered that demand for the firm's product exceeds the capacity decided upon. The decision may, of course, be corrected by building more production units. Error cost may be taken as the mismatch between capacity and demand; gross response time will be the time taken to build new plant so that this mismatch is eliminated, and the cost of new plant will be the control cost. For small production units there may be a whole series of construction programmes which are financially and technically possible. Let these be represented in Table 2.1 where demand exceeds capacity at year 0 by four times the capacity of a small unit, and is expected to continue at this level.

Table 2.1 – Possible Construction Programmes with Small Units

Year	*Number of small units added by end of year*			
1	4	2	1	1
2	0	2	2	1
3	0	0	1	1
4	0	0	0	1

This means that error cost has the pattern of Table 2.2. measured by mismatch between demand and capacity (in terms of the capacity of a small unit).

It can be seen that the family of corrections represented here is well-behaved, gross response time varying with error cost. The corrigibility of the decision to fix capacity at its earlier level with small production units is therefore measured ordinally by [MIN (error cost)]$^{-1}$ = ¼ and [MIN (gross response time)]$^{-1}$ = 1.

For a system of large production units there will necessarily be fewer corrections which can be taken, perhaps only the three in Table 2.3. Error costs may, as before, be represented as in Table 2.4. Again we have a set of well-behaved corrections to the earlier decision, and corrigibility is now $\frac{1}{12}$ when measured by [MIN (error cost)]$^{-1}$ and $\frac{1}{3}$ by [MIN (gross response time)]$^{-1}$.

Table 2.2 – Error Cost for Possible Construction Programmes with Small Units

Year	Error Cost			
1	4	4	4	4
2	0	2	3	3
3	0	0	1	2
4	0	0	0	1
5	0	0	0	0
Total	4	6	8	10
Gross response time (yrs.)	1	2	3	4

Table 2.3 – Possible Construction Programmes with Large Units

Year	Number of large units added by end of year		
1	0	0	0
2	0	0	0
3	1	0	0
4	0	1	0
5	0	0	1

Table 2.4 – Error Cost for Possible Construction Programmes with Large Units

Year	Error Cost		
1	4	4	4
2	4	4	4
3	4	4	4
4	0	4	4
5	0	0	4
6	0	0	0
Total	12	16	20
Gross response time (yrs.)	3	4	5

The corrigibility of the system of small units is, therefore, higher than that of the large production units. In other words, mistakes about the capacity of the system can be remedied more quickly and impose lower costs when small units are employed. Against this, investing in small units is more expensive as scale economies are lost. In other words, control costs are more. In deciding whether to have large or small units, therefore, corrigibility has to be traded against control costs. In this, the decision is typical of ones made under ignorance.

2. The Control of Systems

Making a decision under ignorance can, as we have seen be viewed as being prepared to correct decisions in the light of future information. An equivalent view, which will now be explored, is that making a decision of this kind is to be in control of a system whose behaviour through time is unknown. Any decision can be seen as an attempt to control a system. The above example about the choice of size of production unit can be seen as an attempt to control the system of production with the objective of minimizing the performance indicator, mismatch between capacity and demand. In general, the performance of any system can be represented by a performance indicator, positive when the system is behaving correctly and delivering benefits, negative when it is behaving wrongly and imposing error costs. In Figure 2.2 control is seen as necessary only when the performance indicator reaches some negative threshold T, at time t_M, the monitor's response time. A represents the development of the system if no control is imposed, the area under this curve being the uncontrolled error cost of the system. If a control is imposed, the performance indicator will move along a *control path B*, the area under which is the controlled error cost. t_G is the gross response time of the system, $t_G - t_M$ the corrective response time. The cost of imposing a control is the control cost.

It seems intuitively clear that a system is easy to control if it has a low gross response time and low controlled error cost, i.e. if its misbehaviour is not expensive and can be controlled quickly. Where there are a number of controls, their response time and controlled error cost often vary together, when we can call the set of controls well-behaved. We may then employ either [MIN (gross response time)] $^{-1}$ or [MIN (controlled error cost)] $^{-1}$ as ordinal measures of the system's controllability. In making decisions under ignorance, investment in systems of high controllability should be favoured.

In the example above, for instance, the system of small units is easier to control than the one of large units because imposing a control, i.e. building new capacity, eliminates error costs more quickly. Against this is the loss of scale economies, or an increase in control cost. There is, therefore, a tension between controllability and control costs here, as is typical of decisions under ignorance.

The view that systems of high controllability be favoured in making decisions under ignorance is clearly equivalent to the earlier one that such decisions should be highly corrigible, because identical measures may be used for controllability and corrigibility.

3. Flexibility

Consider now what factors determine the number of options open to the decision maker at t_G in Figure 2.2. At t_{gs} the system may be placed in one of a number of states which are hoped to yield positive performance. The number of such options declines as t_G lengthens. The passage of time, by itself, is a notoriously powerful closer of options. As t_G increases, therefore, options many close, but none may open. There is also a secondary effect. As t_G lengthens, benefits from the system are delayed and so have to be discounted. It follows from the next paragraph that this also tends to diminish the options which are open.

Controlled error cost also affects the number of options which are open. If the aim of controlling the system is long-term profit, then this cost must be recouped by the choice of high performance states after t_G. In general, the system's controller will not know the profitability of each state open to him, but he may be able to rule out some as clearly inadequate. As the controlled error cost rises, the profit required after t_G increases, with more states having to be ruled out as insufficiently profitable. A high controlled error cost, therefore, means fewer options for the system's controller at t_G.

It is customary to talk of the number of options open to a decision maker as the *flexibility* of his decision. What we have shown above, therefore, is that flexibility is inversely related to a system's controlled error cost and gross response time. Where a number of controls exist for a given system, as observed earlier, these two factors will often vary together, and the control is then said to be well-behaved. The flexibility of the system may then be measured ordinally by either [MIN (controlled error cost)]$^{-1}$ or [MIN (gross response time)]$^{-1}$, the familiar measures we

Figure 2.2 *The Control of a System*

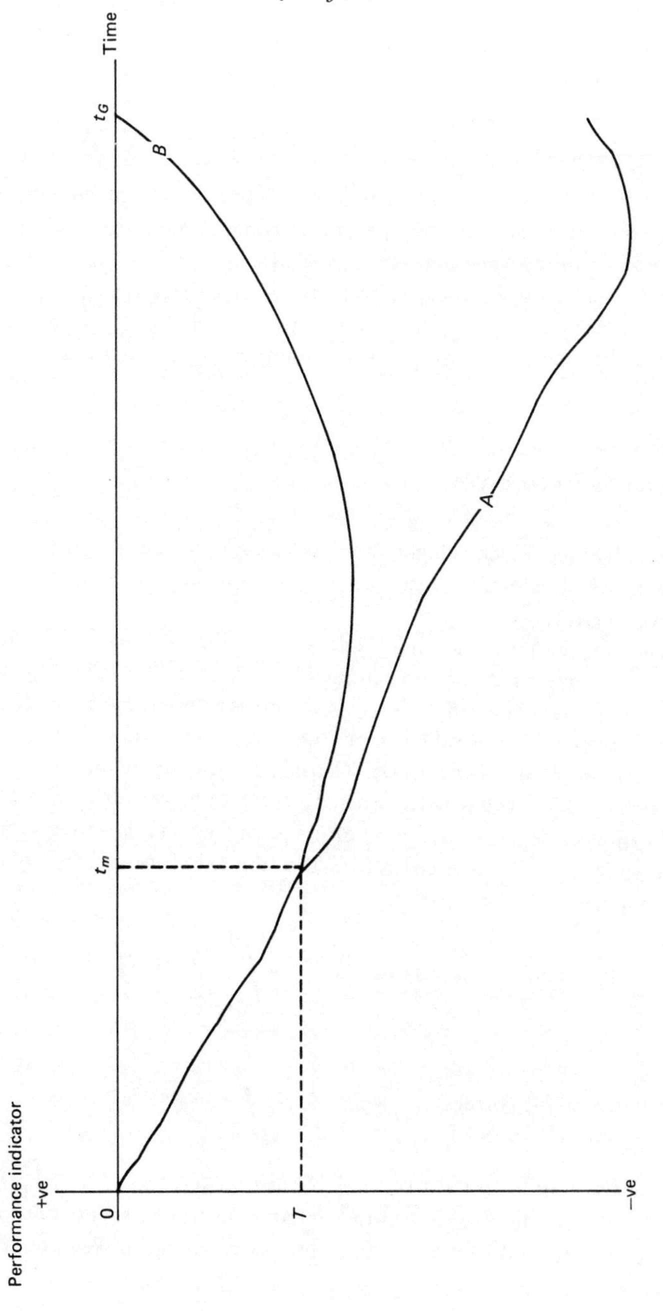

have already used for controllability and corrigibility. This suggests that the art of deciding under ignorance may be viewed in three equivalent ways: make decisions which are easy to correct; choose systems which are easily controlled; keep your future options open. It follows from this equivalence that just as high degrees of controllability and corrigibility usually have to be paid for by high control costs, high flexibility is generally similarly expensive.

The view that decision makers operating under ignorance should keep their options open applies in a particularly straightforward way to the example of choosing between production systems of different unit size. Small units make possible a greater range of capacities than large units; in the earlier example 1, 2, 3, 4 . . . as against 4, 8, 12 . . . , so that the decision maker has more options in deciding the capacity of the whole system.[10]

4. Insensitivity to Error

To make a decision which is highly corrigible or highly flexible or to invest in a system which is easily controlled, is also to make a decision which is insensitive to error.

Consider a decision made under ignorance. If the decision is mistaken, this will be discovered at t_M, and can then be corrected, this taking t_R, the corrective response time. So $t_M + t_R$ is the gross response time, t_G. If the decision is wrong, the controlled error cost which it imposes (a function of time) may be represented by $EC(t)$. The benefit obtained after correction may be written $B'(t)$, the benefits when no correction is needed being $B(t)$. For a decision to be insensitive to error, its payoff when it is wrong should be close to its payoff when correct, so that the existence of error makes little difference to the final payoff.

$$\text{Payoff if decision is mistaken} = \int_{t_g}^{\infty} B'dt - \int_{0}^{t_g} ECdt$$

$$\text{Payoff if correct} = \int_{0}^{\infty} Bdt$$

For the decision to be insensitive, $\int_0^{t_g} ECdt$ must be small and $\int_{t_g}^{\infty} B'dt$ large. The first condition amounts to the requirement of a small controlled error cost. The second condition can be met by reducing t_g. We have, then, our two familiar conditions, low controlled error cost and low gross response time, but now as requirements for a low sensitivity to error.

The above example is simplified because only one correction is considered. In practice, however, there will be many corrections which can be applied to the decision once it has been revealed as mistaken, each with its particular controlled error cost and gross response time. Where these vary together, large cost being accompanied by a long response time, the set is said to be well-behaved. We can then apply the same measures as before, namely [MIN (gross response time)] $^{-1}$ and [MIN (controlled error cost)] $^{-1}$, but this time as ordinal measures of the decision's insensitivity to error. The idea behind the measures is, as before, that they indicate how insensitive a decision can be made to error by a choice of correction. In a typical decision under ignorance there is as before, however, a tension between control cost and insensitivity to error.

The application of this view to the problem of choosing a production system of a certain unit size discussed earlier is particularly straightforward. According to the measures above, the decision to have a system of small units is more insensitive to error than the decision to have one of large units. This is as might be expected because there are always more controls available with the former, i.e. more choice in the capacity added, and it is possible to apply controls earlier than for the large unit system. Whatever the mistakes made in planning the capacity of the whole system, therefore, they will tend to be smaller and more quickly remedied if small units are used rather than large, so that error costs for the system of small units must, in the long term, be lower. As before, this must be traded against the added control cost, i.e. the loss of scale economies.

This concludes the first part of the book's theoretical discussion. It will be resumed in Chapter 10, where it is necessary to take a deeper look at how facts can falsify decisions. For now, however, we are sufficiently well armed theoretically to look, in Part 1, at some historical decisions made under ignorance, all of which, of course, concern the control of various technologies. In looking at history, however, the intention is still the normative one of seeing how decisions about the control of technology ought to be taken. The case studies provide an opportunity to apply the theory developed in this Chapter to the search for obstacles to the control of mature technologies. Decisions about the control of technology have to be taken under ignorance. They ought, therefore, to be made in ways which do not prejudice the future control of the technology; in ways which keep future options open; in ways which make error easy to detect and correct, so that the long-term performance of the technology is insensitive to error. But major technologies simply do not behave like this: as they become more developed and diffused, they become more difficult to control, earlier mistakes become harder to rectify, future options are closed, and decisions become more and more sensitive to error. *Decisions about the control of technology do not seem to be made in a rational way.* Given the damaging social costs which modern technology can inflict, this is a problem of the very greatest importance. We desperately need to understand the origins of the resistance of technologies to control, so that ways of countering it can be discovered and applied in order to improve the quality of decision making in an area so vital to us all.

References

1. Where this is not applicable, rules such as maximin or Laplace's are sometim suggested, but the problems of using these are well known.
2. See, for example, Keeney and Raiffa (1976).
3. von Neumann and Morgenstern (1947),
4. Harsanyi (1953), (1955), (1975); Rawls (1971); for more details see Collingridge (1979).
5. For additional arguments see Collingridge (1979).
6. See, for example, Energy Policy (1978).
7. In Chapter 9.
8. Its principal contemporary champion is Karl Popper. See his (1972).
9. Connolly (1977).
10. For more on the relationship between flexibility and corrigibility/controllability see Collingridge (1979). Other measures for flexibility are suggested in Marschak and Nelson (1962); Merkhoffer (1977); Pye (1978); Rosenhead *et al.* (1972); Rosenhead and Gupta (1968).

Bibliography

D. Collingridge (1979), *The Fallibilist Theory of Value and Its Application to Decision Making*, PhD. Thesis, University of Aston.

T. Connolly (1977), 'Information Processing and Decision Making in Organisations', in B. Straw and G. Salencik (eds.), *New Directions in Organisational Behaviour*, St. Clair Press.

Energy Policy (1978), *Energy Policy – A Consultative Document*, Cmnd. 7101, HMSO.

J. Harsanyi (1953), 'Cardinal Utility in Welfare Economics and in the Theory of Risk Taking', *Journal of Political Economy, 61*, 434 – 5.

(1955), 'Cardinal Welfare, Individualistic Ethics and Interpersonal Comparison of Utility', *Journal of Political Economy, 63*, reprinted in E. Phelps (1973), *Economic Justice*, Penguin 266 – 85.

(1975), 'Non-Linear Social Welfare Functions', *Theory and Decision, 6*, 311 – 30.

R. Keeney and H. Raiffa (1976), *Decisions with Multiple Objectives*, Wiley.

T. Marschak and R. Nelson (1962), 'Flexibility, Uncertainty and Economic Theory', *Metroeconomica, 14*, 42 – 58.

M. Merkhoffer (1977), 'The Value of Information Given Flexibility', *Management Science, 23*, 716 – 27.

J. von Neumann and O. Morgenstern (1947), *Theory of Games and Economic Behavior*, Princeton University Press.

K. Popper (1972), *Objective Knowledge*, Oxford University Press.

R. Pye (1978), 'A Formal Decision Theoretic Approach to Robustness and Flexibility', *Journal of the Operational Research Society, 29*, 215 – 29.

J. Rawls (1971), *A Theory of Justice*, Oxford University Press.

J. Rosenhead and S. Gupta (1968), 'Robustness in Sequential Investment Decisions', *Management Science, 15*, 18 – 29.

J. Rosenhead *et al.* (1972), 'Robustness and Optimality as a Criterion for Optional Choice', *Operational Research Quarterly, 23*, 413 – 31.

PART 1

THE ROOTS OF INFLEXIBILITY

3. ENTRENCHMENT

The purpose of Part 1 is the identification of some of the obstacles to the maintenance of flexibility which has been seen to be essential for the effective control of technology. Just why does a developing technology become more and more resistant to control, and what can be done to counter this tendency? These questions are examined through a number of case studies, the first of which concerns lead in petrol and highlights what may be called the *entrenchment* of technology. This refers to the adjustment of other technologies to one which is developing, so that eventually control of the latter is only possible at the cost of re-adjusting the technologies which surround it. Control has therefore become difficult, expensive and slow.

Case Study – Lead in Petrol

This case study illustrates the concept of technological entrenchment. It concerns lead additives in petrol and the difficulties in controlling lead from this source. Lead has been added to petrol in the United Kingdom since 1923, and a few years later was used throughout the world. At this time concern about the health effects of the additives centred on those occupationally exposed – policemen, garage attendants and so on. The amount of lead spread over the country by motor-car exhausts was not regarded as a hazard, although the enormous increase in the car population and its mileage was not then foreseen. If it were shown that lead emitted in this way is a hazard to the health of the general population, then the original decision to add lead to petrol would be regarded as wrong. Monitoring the decision is, therefore, the search for damage to health from this source of lead. The present case study considers the options open if monitoring were to reveal such damage.

Thermodynamics tells us that the thermal efficiency of a gas engine increases as the compression ratio of the working gas is increased. If the efficiency of a given engine is increased by increasing its compression ratio, then it will consume less fuel for the same output of work, or else the output will be produced more quickly, giving greater power. Engine designers, therefore, place a premium on high compression ratios. For the petrol engine compression ratios are limited by the phenomenon known as knock (or pinking). The mixture of air and petrol in the cylinder is compressed and then ignited by a spark from a sparking plug. The mixture should burn evenly so that the flame front from the plug expands smoothly. Knocking occurs when some of the mixture before the flame front becomes sufficiently hot to ignite spontaneously. This gives uneven pressure on the descending piston which can sometimes cause considerable damage to the engine.

The response to this problem has been the development of petrol mixtures with a low tendency to spontaneous ignition. With such fuels, high compression ratios can be achieved without knocking. The resistance of a fuel to knock is measured by its octane rating. A standard engine is run on the petrol to be tested, and its compression ratio increased until knocking just occurs. The engine is then run at this compression ratio on a mixture of two hydrocarbons, *n*-heptane, which is very prone to cause knock, and *iso*-octane, which is very knock resistant. The ratio of the two hydrocarbons is adjusted until knocking just occurs. The proportion of *iso*-octane in the mixture is the petrol's octane number. Running the engine under fairly mild conditions gives the fuel's *research* octane number (RON); under more severe conditions the fuel's *motor* octane number (MON).

The production of high octane petrol is expensive, so the motorist pays more per gallon than for petrol of lower octane number. But each gallon enables him to travel further or to use an engine of greater power, by allowing the use of high compression ratios. The optimum octane number is that which achieves balance between low fuel consumption or high power and the cost of fuel. As the real extra cost of high octane petrol has fallen over the years, compression ratios have increased as shown in Figure 3.1.[1]

Early in the motor car's development in the 1920s, it was discovered that the addition of small amounts of organo-lead compounds, notably lead tetramethyl and tetraethyl, could bring quite dramatic increases in a petrol's octane number (of the order of 10), and, until recently, all petrol throughout the world was improved in this way. There is now, however, growing concern about the health hazard of the lead from petrol which is emitted from car exhausts, and strong movements exist in many

Figure 3.1 *Average Compression Ratio of New Vehicles*

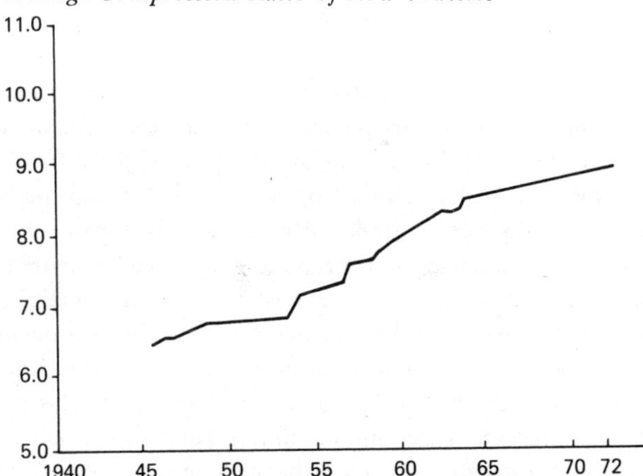

countries for severe reduction in the amount of lead added to petrol, or for its removal altogether. Lead-free petrol is now available in the United States, but the problems posed by its removal there are unique to that country, owing to the way in which oil is refined there, which is quite different from the refining pattern of Western Europe. The discussion which follows, therefore, is confined to Western Europe.

In Western Europe, there has been a protracted, and often bitter, debate about the removal of lead from petrol in nearly all countries and large research budgets have been disposed of, by governments, refiners and additive manufacturers, to further the debate. To say that all this fuss is because reducing lead levels in petrol will be a very expensive business tells us little. What is much more informative is to determine *why* it will be such a costly operation. I suggest that this is because of the way in which the technology of petrol production has interacted with the technology of car manufacture, so that the production of leaded petrol has become entrenched.

The car population of Western Europe represents an enormous fixed investment which is depreciating relatively slowly; the number of new cars sold each year is about 10 per cent of the existing stock. Most of these cars have been designed to run on high octane (four-star or premium) fuel, and attempting to run them on anything else will involve large costs in extra petrol consumption, reduced power and engine damage. To produce this fuel requires sophisticated processing of crude oil fractions and the addition of lead to boost octane numbers. If cars are to continue running on high octane fuel, then any severe reduction in the amount of lead added will require major modification of existing refineries which will be very expensive and which will take several years to complete. To see this in detail unfortunately requires a few more technicalities.

Let us begin with the car. About 70 per cent of European cars are designed to run on high (98 RON) octane fuel (about 75 per cent in the United Kingdom). If it were possible for these cars to run on lower octane petrol, then reduction of lead would pose little difficulty. Existing refineries would merely refrain from adding lead, except perhaps in small amounts, and the petrol produced would still be useable in the majority of cars. Fuel economy would suffer, of course, and so more petrol would be consumed, but such is the cost of environmental improvement. Unhappily, however, running engines designed for high octane petrol on petrol of lower octane number is enormously expensive. Knocking could produce serious engine damage on a very large scale. This could be mitigated to some extent by ignition timing adjustments and lean running, but this is very expensive in terms of fuel consumption and loss of power, and, in engines not designed for these adjustments, expensive problems such as overheating and increases in noxious emissions would occur. The costs of failing to provide a petrol which can satisfy the vehicle population are, therefore, very large. Figure 3.2 gives the octane number of fuel needed to satisfy varying percentages of the 1968 European car population. The curve has changed very little since then.[2]

Figure 3.2 *Octane Requirements of the 1968 European Vehicle Population*

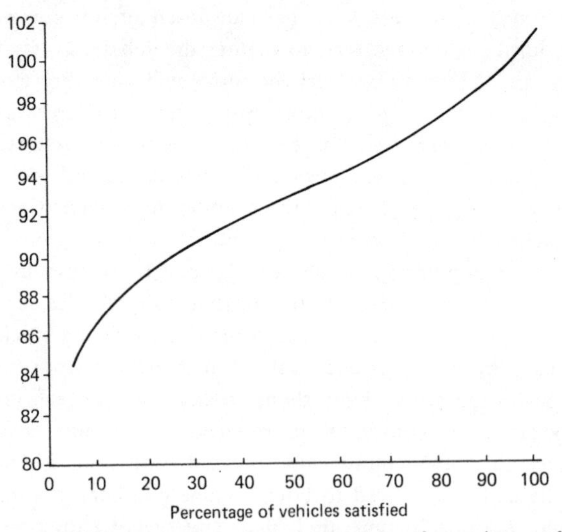

The obvious way of introducing engines capable of running on lower octane fuel than existing ones is to insist that all new cars are fitted with engines of the new design. In about ten years all cars would then be capable of using low octane fuel and the need for lead additives would have completely disappeared. But what if the health hazard from lead emissions is reckoned so great that its elimination in a shorter period, say five years, is thought desirable? If all cars are to run on lead-free fuel in five years time, then the rate at which engines are produced must suddenly be doubled for this period; half the output being fitted to new cars and half replacing engines of the earlier design in old cars. This would be enormously expensive. There is, first, the investment in new engine production plant. This must be reckoned as much greater than the investment made in the plant existing at the moment, even though they would be of the same size, because the new investment has an extremely haphazard lifetime. After five years, the new plant would have to close because only engines for new vehicles would then be needed. But, assuming a ten year life for vehicles, cars whose engines have been replaced will begin to wear out after the new plant has been closed for five years. It will therefore need to open again for five years, after which it will close for five years and so on. To this must be added the cost of fitting the new engines and scrapping the old ones before they are worn out.

What this means is that any reduction in the ten years allocated for the introduction of the new engine able to use low octane fuel is hugely expensive, and as the introduction period is reduced the cost rises dramatically. There is an introduction period, for the sake of argument let it be three years, which is simply impossible because new plant cannot be constructed in time, and we may count the cost of this choice of introduction period as infinite.

It may help to see this in terms of the theory of decision making under ignorance. Error is the imposition of ill health on the general population through their exposure to lead emissions. The error cost is the cost (both tangible and intangible) of this impairment of health. Control cost is the cost of replacing old engines by new, and corrective response time is the period chosen for the introduction of the new engine. The relationship between these factors is shown in Figure 3.3. This reveals the tension between control costs and error costs. Notice that the error cost associated with each control path varies as the control path's response time, so the set of controls is well-behaved. The measures developed earlier can, therefore, be used in estimating the flexibility of the car population to engine changes.

In practice, control costs rise so steeply as response time is reduced, that the only feasible option is the introduction of the new engine over a ten-year period. The system seems to show a 'natural' rate of change; any attempt to increase the rate proves to impose enormous costs.

Figure 3.3 *Control on Environmental Lead from Changing Engines*

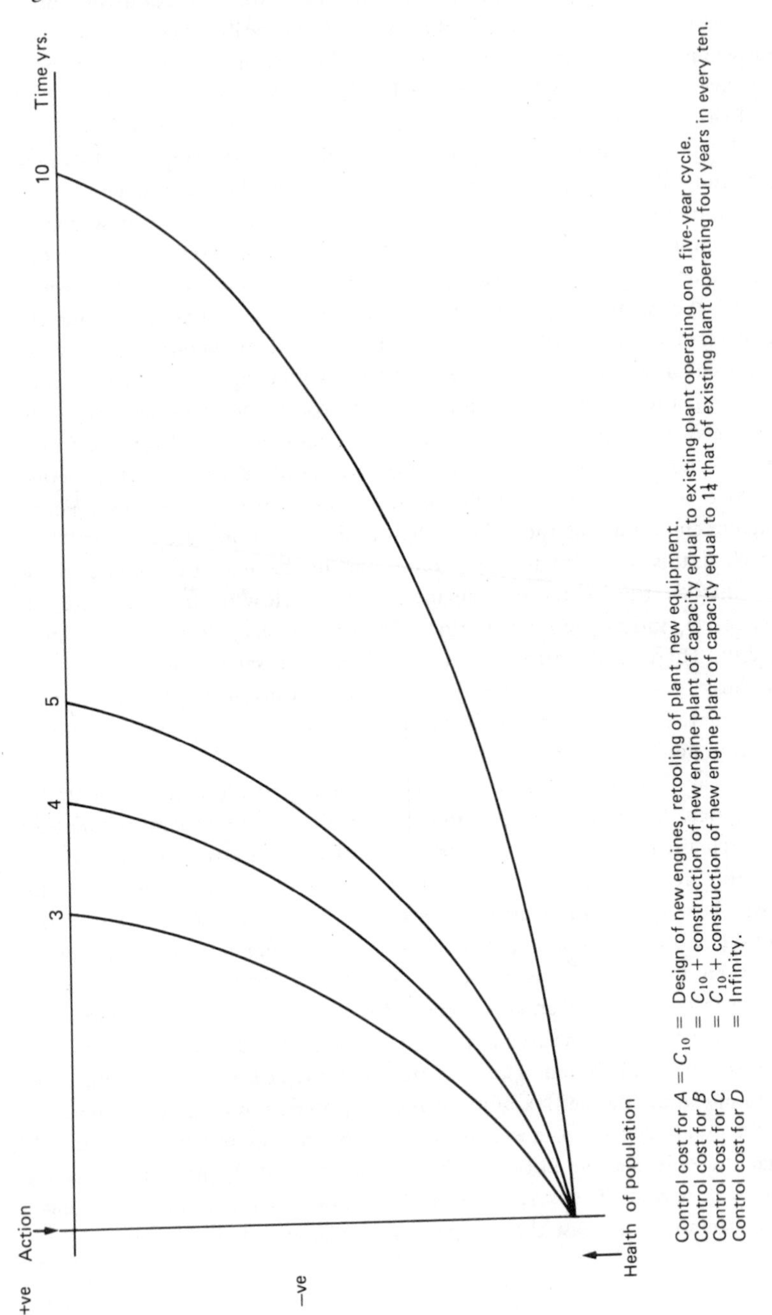

This inflexibility in the control of motor-car engines is a consequence of the entrenchment of the motor car which has already been mentioned in Chapter 1. There it was observed how alternative methods of transport, diesel engined buses and passenger trains, had accommodated to the existence of the motor car. It is, therefore, impossible to make a sudden shift from cars to these other forms of transport; there is simply not enough capacity, nor enough production plant to increase capacity quickly. This is particularly true of rail transport where new lines would have to be laid to allow for the extra traffic. A sudden shift from cars to buses and trains would be impossible, even if the running stock were available, because the split between petrol and diesel which is made at the refinery is made to satisfy the present distribution of transport between cars and buses and trains, and this cannot be changed quickly. There would simply not be enough diesel fuel.

If the hazard from lead in petrol is reckoned great enough to warrant its urgent elimination, and if this is to be done by changing to engines of lower compression ratio, one way of avoiding the massive investment in production plant of erratic life discussed above would be to shift the whole of the motor-car transport to buses and trains, allowing back one-tenth of the traffic per year as cars with the new lower compression engines became available. But the adjustment of these other transport technologies to the existing motor-car population makes this impossible. In the same way, the simplest solution of just foregoing car journeys until the new engines are available is made enormously expensive because people have adjusted their economic and social lives to motor-car transport as witnessed by, for example, the growth of satellite villages. People could leave their satellite villages and move into cities to be closer to their work, but not at all quickly, and at huge cost from building more houses in cities, or forcing people to live in crowded accommodation. In this way all the options to the steady substitution of new engines over ten years are hugely expensive.

The transport system is an example of a highly valuable technological system of lower variety. The best measure of value is perhaps to imagine the cost of the system's total failure. Low variety means that there are only a few ways of performing the function of the system, in this case moving people and goods. Of foot, bicycle, canal barge, motor car, bus, train, ship and aeroplane, the first three are of little importance, as are the last two for local journeys, so the effective variety for these is car, bus and train. These methods of transport obviously reach a mutual accommodation, as do interacting technologies, such as car production, fuel refining and so on. Any sudden and serious disruption of one method is therefore very damaging, because the adjustments required to this accommodation are many and costly and have to be done slowly, as we have seen. If there were fifty independent ways of travelling, then a disruption of one would mean only minor adjustments to the remaining forty-nine. This teaches us to look very closely at similar highly valuable, low variety systems, some more of which will be examined later. Figure 3.4 gives some idea of the scale of the problems of rapidly shifting from road to bus and rail transport.[3]

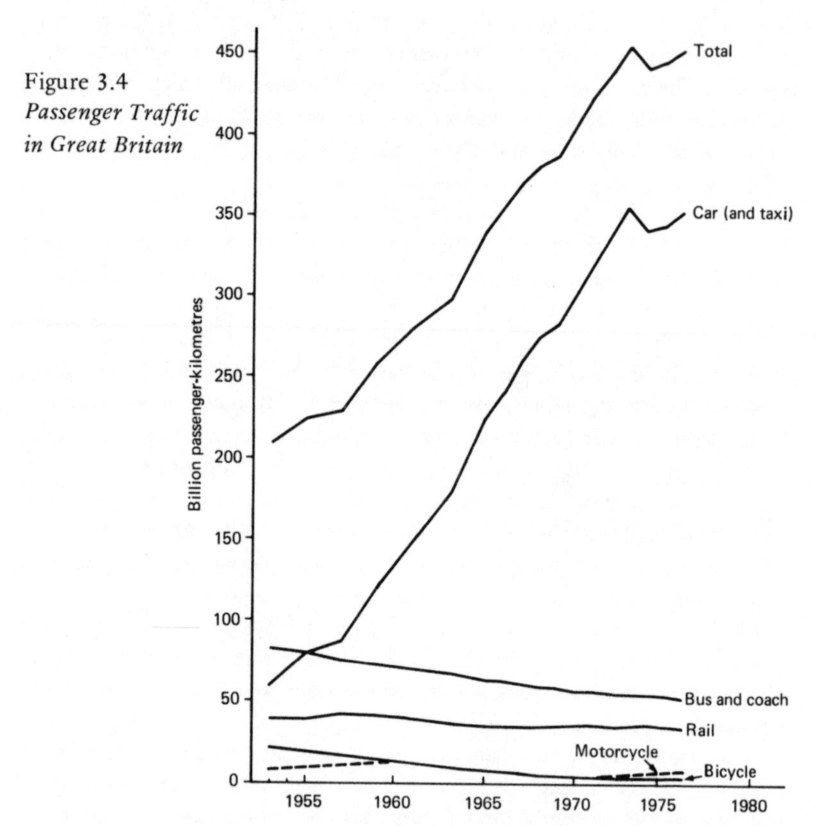

Figure 3.4
Passenger Traffic in Great Britain

Now let us turn to petrol. If high octane fuel could be produced by existing refineries without adding lead, then the problem of reducing lead levels in petrol would be very simple, but unfortunately this is not the case. Producing an adequate high octane petrol is difficult because knock is a slightly more complicated affair than what I have so far said might suggest. There are really two kinds of knock, one which occurs at low engine speed and one which happens at high speed or open throttle acceleration. To determine a petrol's resistance to both kinds of knocking, we need three octane numbers. The *research* octane number (RON) is relevant to low speed knock, whilst the *motor* octane number (MON) and the $RON_{100°C}$ (the RON of that fraction of the petrol which boils below 100°C) are relevant to high speed knock. A typical British four-star petrol, required by about 75 per cent of British cars, has a RON of 98, and MON of around 88 and a $RON_{100°C}$ of around 89. To satisfy these requirements and all the other requirements placed upon modern petrol (high enough volatility to prevent vapour lock, but low enough to ensure rapid warm up; low gum formation; low engine deposit formation; the correct distribution of components to give good fuel economy; low crankcase oil dilution etc.) is very demanding for the refiner.

In Western Europe about 15 per cent of a barrel of oil ends up as petrol. The refiner selects various fractions from the fractionation of the crude oil for processing to petrol. Only a very small part of modern petrol is simply an unprocessed crude fraction (straight run gasoline). The major processes used to convert these fractions to mixtures which will be finally blended to give a petrol are catalytic cracking (of napthas), catalytic reforming (of naptha or gasoline), and isomerization (*n*-paraffins to higher octane *iso*-paraffins). Nearly all Western European refineries have reforming plant and about half have catalytic cracking plant. The various products of these processes are finally blended, along with lead additives, to give a petrol of the required properties.

If petrol of existing high octane quality is to be produced using less lead than at the moment, the Western European refining system has only a limited flexibility to accomplish this. In West Germany, for instance, legislation limits lead to 0.15 g/1, a limit which poses German refineries with the problem of providing a petrol with an adequate $RON_{100°C}$. This is achieved by importing surplus hydrocarbons from European refineries, but the German change and reductions by other European countries have virtually exhausted available output of these hydrocarbons. If lead levels are severely reduced on a large scale elsewhere, existing refinery plant will not be able to produce petrol of an adequate $RON_{100°C}$ (except at the cost of deficient MON or overall RON). In order to meet the demands of a low lead or lead-free, high octane petrol, the refineries of Western Europe must invest considerably in additional processing plant, particularly of the kinds mentioned above.

Such large investments cannot be made overnight. A great deal of co-ordinated planning by the refiners is required, which takes time, as does obtaining planning consents for construction. In addition, the existing plant building industry cannot build above a certain speed and so forms a bottleneck for any rapid expansion. Expansion may also be limited by problems in acquiring the necessary capital. Large chemical plants take at least two years to build, and planning delays can add two years. With an additional year's delay to allow for the construction bottleneck, it may take refiners at least five years before they can produce low lead, high octane petrol; even allowing no time for consultation between the legislators and refiners.

Table 3.1 – Cost of Reducing Lead Levels in Petrol

*Maximum level of lead in petrol g/1**	*Cost 1976 $ x 10⁶*
0.40	260 – 670
0.15	2,400 – 3,600
0.0	15,000

* Current maximum is 0.45 g/1

Apart from the investment cost mentioned above, annual costs will be incurred by the refiners because more processing of crude oil will be necessary before it can be converted into petrol. Estimates, by the refiners, of total costs for various maximum levels of lead in petrol are given in Table 3.1.[4]

So far we have looked at the two extreme responses to the call to reduce lead levels in petrol; altering engine design to enable the use of whatever octane fuel present refineries can produce without adding as much lead as now, and adjusting refineries to produce low lead or lead-free high octane petrol which can be used in car engines of existing designs. In between, of course, are a great number of options where both engine design and refining patterns are adjusted. Refineries could turn to producing low lead or lead-free petrol of octane rating between that of existing petrol and the rating of existing unleaded petrol. Car engines would then need to be modified, but less than before. This mutual change is almost certain to follow any legislation to severely limit lead additives. It does, however, involve the same problems of timing as before. Existing cars will require high octane petrol, presumably achieved by adding lead, and only new cars could run on the low lead or lead-free petrol. As before, this means that the problem will only be finally solved in ten years, the time taken for a more or less complete change in the car population.

Economic forces will drive such a mutual adjustment of engine design and fuel, just as they have produced the existing match between powerful, high compression engines and leaded, high octane petrol. Restrictions on lead additives will make high octane petrol considerably more expensive than lower octane fuel, and so there will be a tendency for motorists to favour engines which are either less powerful or have a higher fuel consumption but which can run on cheaper, lower octane fuel. In this way, the market will generate a new optimum trade-off between power or fuel economy and petrol price. The achievement of this optimum will, however, be delayed because much of the R & D work which will be involved in the mutual adjustment has not yet been done. This may become clearer from considering an example. Car manufacturers only invest in those R & D programmes which may lead to improvement in their products which are significant enough to have some impact on their market share or profitability. It is well known that engine designs vary enormously in their proneness to knock. Engines of the same compression ratio can require fuel which differs by as much as ten octane numbers. It would appear however that little understanding of the reasons behind this variation has been acquired. The reason is not hard to see. With the very small price difference between regular (two-star) and premium (four-star) petrol, getting an engine to run on the lower grade fuel is not very significant at the market place, and probably not worth the considerable R & D costs involved.

This changes, however, once the price difference between the fuels increases, as it will if lead additives are restricted. It may then appear very worthwhile to invest R & D effort in understanding why engines of the same compression ratio can vary so widely in their octane requirements. But it is likely to be some time before this knowledge is acquired. Reaching a new optimum balance between fuel costs and engine power and fuel consumption will, therefore, be delayed.

All of the difficulties we have discussed about reducing lead levels in petrol disappear if a workable lead trap, a device for filtering out lead from the exhaust, becomes available. Existing engine design and refining patterns can then be left unaltered, and environmental lead levels reduced by the simple device of inserting such a trap in the exhaust system. This will be expensive, of course, but environmental improvements must be paid for. This solution has no timing problems either since traps can be fitted to existing cars. It is hardly surprising, therefore, to find out that a great deal of reasearch into lead traps is going on, funded by the lead additive manufacturers, car manufacturers and oil companies. Nor is it surprising to find that considerable prominence is given to the development of the trap in arguments for the retention of the status quo. Traps offer a simple way of leaving engine technology and refining technology locked together as they are, thus avoiding the costs of separation.

Lessons from the Case Study

(a) Entrenchment means that change is expensive

We have seen that because the technologies of fuel production and car production have become locked together, a change in one, such as severe reductions in lead additivies, inevitably requires change in the other. Imagine that the two technologies were independent and that the present car population could run on the lead-free products of existing refineries, or else that existing refineries could produce high octane petrol without adding lead. Then the removal of lead might well involve recurrent costs; as using unleaded fuel might be more expensive than using the leaded product, but only small capital costs would be involved. As it is, rapid and severe reduction in lead additives will require one or both of major capital investment in refining plant and costs from damaged vehicles running on inadequate fuel.

(b) Entrenchment means that the timing of change is fixed

In reviewing the problem of reducing lead additives in petrol, one cannot but be struck by the impression that the timing of change is determined more by the technology itself than by human intervention. Any attempt to reduce lead levels quickly, in less than five or six years, involves very high costs. Refining patterns cannot be altered in less than this time, so removing lead tomorrow would mean bearing the enormous costs of five or six years of cars damaged from being run on inadequate fuel, or the equally huge costs of foregone car journeys. Similarly, altering engine design over a period of less than about ten years involves huge expenditure on short lived car plants and the costs of prematurely scrapped vehicles.

(c) Entrenchment means that change is hotly debated

Why is the reduction of lead additives in petrol so contentious an issue? Surely, because such a reduction will be very expensive, but it will be expensive, and also protracted, because of the way in which lead additive technology has become entrenched. If it were not for this entrenchment, the reduction of lead levels would be so straightforward an issue that debate on the scale which we see today would be quite superfluous. I would suggest that under any continuing and hot debate about technological change, a little investigation will unearth a technology entrenched in the same way as lead technology.

(d) Entrenchment means that in the debate about change, the status quo has an unfair advantage

In talking to informed proponents of reduction in petrol lead levels, one soon senses the frustration which they feel. They think that they have made out a good case against lead in petrol, and yet action seems very far away. Let one remark stand for many – 'if you can't pin it on lead in petrol, what can you pin it on?'. In a world very much tidier than the real one, the scientific task of determining whether or not lead from petrol is harmful would be separated from the political one of deciding how much money to spend on remedying the situation. In our more chaotic world, however, these two aspects of the problem become conflated. Because the reduction of lead levels in petrol will be very costly and protracted, the *scientific* case against lead has to be made very much stronger before being accepted than would otherwise be. This will be discussed more fully in Part 2.

(e) Entrenchment means that achieving a new optimum adjustment of technologies is a lengthy procedure

The consumers' trade-off between power and fuel economy and petrol cost determines the point at which the two technologies of engine and petrol production became locked together. Breaking this lock by reducing the refiners' dependence on lead additives, means that a new optimum situation has to be found eventually. As we have seen, reaching the new optimum makes new demands on R & D efforts and so is likely to take a long time.

(f) Entrenchment means that 'fixes' are strongly favoured

If some way can be found of alleviating the problem which gives rise to the call for change in an entrenched technology without actually making the change, it is obviously going to be received very favourably. We have seen this with the lead trap which by-passes the problems involved in altering the relationship between car and fuel production technologies.

(g) Entrenchment is particularly severe for highly valuable, low variety systems

Two such systems will receive attention in Chapter 5, the energy system and the peace keeping system.

It is apparent that all of these features apply not just to the present case, but to all cases of entrenched technologies. In controlling technology, therefore, entrenchment is a very serious obstacle. This raises the question of whether entrenchment on a scale sufficient to be a serious obstacle to control can be predicted, or can it only be identified once it has happened; and the question of its avoidance. We shall see in Chapter 9 that it is sometimes easy to foresee that a particular technology will become highly entrenched if its development is not controlled at an early stage. The chapter also considers ways of tackling the danger of entrenchment. As noted above, entrenchment is particularly harmful in highly valuable, low variety technological systems, so that one way of avoiding it is to insist that variety is increased by the development of several technologies which perform the same function. For energy technologies, this amounts to such things as insisting on technologies which use a number of primary fuels, and the development of different ways of supplying delivered energy such as liquid fuel. In the case of the motor car, variety could be increased by developing other forms of transport on a comparable scale to the car.

Another way to avoid entrenchment, or at least to alleviate it, is to make relatively minor changes in hardware. As observed above, the high compression petrol engine is extremely fussy about the fuel it is given. It is reasonably certain, however, that oil will become increasingly scarce and expensive and that some alternative source of liquid fuel will be required. With a huge fixed investment of high compression engines, the demands made on the quality of this fuel will be very severe, and the penalties for failure to meet them will be large. As change is so slow here, it would make the switch to the new liquid fuel much simpler if engines were of lower compression ratios and so able to tolerate poorer fuel. This would have to be done soon if a significant change in the engine population is to have occurred before the expected substitution becomes necessary. An analogy to the sort of action necessary is the American legislation which lays down minimum averages for the petrol consumption of motor cars of various kinds.

References

1 Ellis *et al*. (1973).
2 Concawe (1972).
3 From Leach (1979), p. 135.
4 Collingridge and McEvoy (1980).

Bibliography

Associated Octel Co. Ltd. (1977), *The Cost of Control of Lead Emissions: Lead Reduction vs. Exhaust Gas Filters*.

D. Collingridge and J. McEvoy, 'The Cost Effective Comparison of Controls on Environmental Lead', Technology Policy Unit Working Paper, forthcoming, *International Journal of Enivonmental Science*.

D. Collingridge (1979), 'The Entrenchment of Technology – The Case of Lead Petrol Additives', *Science and Public Policy, 6*, 332-8.

CONCAWE (1972), *The Problem of Gasoline Engine Exhaust Control*, Report 12/72, The Hague.

CONCAWE (1977), *Automotive Emission Regulations and their Impact on Refinery Operations*, The Hague.

D. Ellis *et al.* (1973), 'The Environment, Petrol, the Car and the Cost', in *Passenger Car Engines*, Institute of Mechanical Engineers, London.

Ethyl Corporation, *Particulate Traps*, Ferndale Michigan, 1977.

D. Leach, *et al.* (1979), *A Low Energy Strategy for the UK*, Science Reviews.

K. Owen (1973), 'Motor Gasoline' in G. Hobson and W. Pohl (eds.), *Modern Petroleum Technology*, 4th edn., Applied Science Publishers, 573–613.

US Environmental Protection Agency (1979), *Control Techniques for Lead Air Emissions*, (EPA-450/2-77-012), North Carolina.

4. COMPETITION

The greater part of the world's technology serves a competitive function; be it in economic competition between private firms or whole countries, or in military competition. It is, therefore, of some importance to examine the effect of competition on the quality of decisions about the control of technology. This is especially pressing when it is remembered that all important decisions of this kind have to be made under ignorance, when it would seem that competition inevitably leads to bad decision making. The essence of making decisions under ignorance is to keep future options open, but the existence of competition necessarily reduces the number of options which may be taken up. The need for the chosen technology to be competitive with the technology of rivals places an added restriction on the selection of technology, and so further limits the choice which is available.

A firm might, for example, be happy to make their product in a number of ways which employ different technologies, especially when it is remembered what a simplification it is to see the firm as a profit maximizer. The firm may well trade off some of its profit with its desire to keep together a good workforce, to avoid merging with other companies, to avoid troublesome investment in new equipment and so on, so that a large number of options exist for the firm to produce its future output. If it is now demanded that the firm be profitable in a competitive market, the number of ways it can decide to produce its output is immediately reduced. If the firm's knowledge about future markets, raw material prices, labour relations, the effects of technological innovation on its own and its rivals' goods, and so on, is insufficient for the optimal decision to be made, then the decision is under ignorance. The reduction in options must then be viewed as bad, as it reduces the control the firm has over its technology and reduces its flexibility.

Another way of making the same point is to say that competition increases the sensitivity of the payoff from decisions to error. If a non-competing firm makes a mistake in its choice of technology, so that its product is more expensive than it needs to be, then the cost of this error is the loss of profit which results. If the same mistake is made in a competitive market, then its cost is greatly increased. One of the firm's competitors is sure to avoid this error, and so sure to be able to obtain a cost advantage which may lead to a severe reduction in the firm's profits, or even to its bankruptcy. But we saw in Chapter 2 that decisions under ignorance ought to be made in such a way as to be insensitive to error. As before, competition leads to bad decision making.

This is bad enough, but where the technology involved in the competition has a significant lead time the consequences can be much worse. If *X* and *Y* are engaged in a competition of this sort, a pattern of decision making may arise of the following form.

(i) Competition increases error costs. It is therefore essential for *X* to avoid the very high error cost which would result from *Y*'s adoption of technology *T*, and *X*'s failure to adopt *T*.
(ii) *X* cannot postpone his decision about the adoption of *T* until he discovers that *Y* has adopted it, because it takes a long time to develop *T*, during which very large error costs will accrue.
(iii) *X* therefore decides now to develop *T*, to prevent *Y*'s unilateral adoption of it.
(iv) *Y* argues in a symmetrical way, and so also decides to develop *T* now.

In this way both parties are forced to adopt the most competitive technology, whatever they may think of its merits. They may both regard the technology as socially damaging, but provided it is less damaging than to lose the competition, they have no choice but to adopt it. The existence of competition forces choices which are hedges against the worst outcome, rather than positive decisions. In such a situation it is quite wrong to talk of choice at all. The flexibility of the decision is zero; there is no control over the technology; payoff from decisions is extremely sensitive to error; and error is very hard to correct, all of which is attributable to the exsitence of competition. Competition involving technologies having long lead times is extremely common and provides one of the most pernicious sources of the resistance of technologies to control. The madness of the decision pattern above must be avoided if we are ever to be able to pretend that we exercise control over our technological creations.

An example which we are living through at the moment is microelectronics. As mentioned in Chapter 1, the spread of this technology may lead to very severe unemployment. This is, however, a cost which does not figure in the market price of the goods and services provided by the technology. The British Government's thinking, and the consensus elsewhere, is that the United Kingdom has no choice but to adopt and to press ahead with the diffusion of microelectronics because, whatever the unemployment consequences, the cost of not having this technology when it is in the hands of our economic competitors is far greater. There is no choice, no control, no flexibility – the United Kingdom has to have it. Thus the National Computing Centre tells us that the country has 'no choice, no control, no flexibility – the United Kingdom simply has to have it. Thus the National Computing Centre tells us that the country has 'no choice but to go forward with microelectronics technology or get out of business',[1] and Braun declares that 'what options we have arise out about the adoption of the technology, whatever the social costs it inflicts. Thus is found the absurd prospect of no competitor wanting microelectronics because of the social damage it may do, but each being forced to have it as a hedge against even worse costs.

The case study which follows is simple in two ways. There are just two competitors, the superpowers, not the dozens or hundreds of commercial or international competition; and the nature of the competition is particularly straightforward, being the military conflict between them. The example is the nuclear weapons technology known as MIRV.

Case Study – MIRV

This case study examines the decision to develop and deploy the military technology known as MIRV (multiple, independently targetable re-entry vehicle), from the standpoint of the theory of decision making of Chapter 2. An excellent history of MIRV is given in some detail in Greenwood (1975) and most of the present Chapter is based on his account. The key question for us is whether the decision to develop and deploy MIRV was a rational one, given the view of decision making under ignorance developed earlier and, if it is found not to be, why was this so. To make rational decisions in a state of ignorance is to be in control of a system; it is to be able to acquire information about the system's behaviour and environment and to be able to modify the system's behaviour accordingly. Our question can, therefore, be re-phrased as 'was the development of MIRV under the control of the relevant decision makers?'. Of particular concern will be the effect of competition on the ability of these decision makers to control MIRV's development.

Early ballistic missiles carried one nuclear warhead protected from its journey out of the atmosphere and back again by a single re-entry vehicle.

A missile fitted with MIRV (or a MIRVd missile) carries a number of warheads, each protected by a re-entry vehicle, on a bus (Post Boost Control System). The bus is carried by the missile's final stage and is fitted with a guidance and control system which can change the bus's velocity, orientation and trajectory using small rockets. The missile booster places the bus in a trajectory close to that required for the first re-entry vehicle. The trajectory is then adjusted after the booster has fallen away and the first re-entry vehicle with its warhead is dropped from the bus. The flight of the bus is then adjusted and the second re-entry vehicle dropped and so on. Some re-entry vehicles can carry penetration aids in place of a warhead, such as decoys to attract opposing fire or chaff to confuse enemy radar. MIRV was developed in the United States in two forms, the Mark 12, carrying up to three warheads, for fitting to the Air Force's landbased Minuteman missiles, and a version fitted to the Navy's submarine-launched Poseidon missiles and carrying up to fourteen warheads. One of Minuteman III's warheads has an explosive power of around 200 kilotons compared to the 50 kiloton warhead of Poseidon (the bombs dropped on Japan had a power of about 20 kilotons, i.e. equivalent to the explosion of 20,000 tons of TNT).

MIRV was a straightforward, though expensive, extension of existing technology and work on all its components had been undertaken, either in the US Space Programme or in weapons development. For this reason it is not surprising to find MIRV is an example of multiple invention, no less than five groups sharing this honour in late 1962 and early 1963. It was very much a product of the technical community and there was no definite outside military interest pressing for its development. The key decisions to develop MIRV were made by the Secretary of Defense Robert McNamara, for Poseidon in late 1964 and for retrofitting to the Minuteman fleet in the following year. The decision here was to develop the technology of MIRV, not to deploy it, which called for another round of decisions. Development, however, went forward very smoothly and deployment soon followed. The Air Force took operational control of the first ten Minuteman III fitted with MIRV in June 1970, and the following August the first submarine launching of the MIRVd Poseidon occurred.

The development of MIRV was surprisingly smooth and unruffled. In Greenwood's words:

> The most striking characteristic of the process that led to the MIRV innovation is the speed and ease with which it operated. The programme proceeded about as fast as technology would allow. This is not because the technical community was given free rein to do what it

> pleased, but rather because all relevant centres of authority, including the Secretary of Defense, rapidly came to think that MIRV was a desirable system. Those who did not totally share this belief were soon won over, like the operational side of the Air Force, or were excluded from the decision-making process, like the Arms Control and Disarmament Agency. In its formative years MIRV was a programme that contributed to the objectives of all organizations and individual decision-makers that participated in the innovation process.[3]

For our purposes there is little benefit in a detailed description of the various actors and groups involved in the decision-making process, all of which can be found in Greenwood's book. They will emerge as far as is needed in order to assess the rationality of the decision. Before this can be done, however, it must first be shown that the decisions to develop and deploy MIRV were made under ignorance.

The objective of American strategic planning is to deter the Soviet Union from launching a nuclear attack against the United States. To do this the United States needs to be assured that following an attack it can inflict damage on the Soviet Union on such a scale that it would never pay that country to attack the United States. This ability is obviously a function of the military technology deployed by the United States and by the Soviet Union. In planning this technology, therefore, the United States would ideally like to know everything about the technology of its potential opponent. This has led to an extensive intelligence gathering exercise, the scale of which increased enormously in the years just before the invention of MIRV. Defence decision makers in the United States since this time have been inundated with intelligence, but despite this, they must still operate under the greatest ignorance. Spy satellites can tell where a missile landed, but not where it was aimed, so it is very hard to determine one of the crucial characteristics of missiles – their accuracy. Similarly defence systems are very difficult to assess, e.g. the capabilities of a new radar system cannot be inferred from its shape, size and position.

These, and similar things, are inherently difficult targets for intelligence gathering, but even if information could be obtained it could reveal nothing about Soviet intentions. Military intelligence may give the United States everything it wishes to know about what its opponent *can* do, but it can tell little about what it *wants to do*. This puts military planners in a very difficult position, and so for reasons of caution they tend to act on Soviet capabilities, i.e. on estimates of the worst they could do, and not Soviet intentions. This is reinforced by the fact that intentions can change in months or even days, whereas military capabilities can only change over several years as new technology is developed and deployed. This is, of course, one way of hedging against ignorance about the present and future intentions of one's opponent.

Knowing what capabilities the Soviet Union has would enable the United States to develop weapons to ensure that deterrence is maintained, but weapons take a long time – typically five to ten years – to develop and deploy. Ideally, therefore, the United States should know what weapons its opponents will have deployed in, say, ten years' time, so that it can design counterweapons and develop and deploy them when they are needed. But such projections of Soviet developments are extremely difficult. McNamara himself recognizes this in his FY 1965 Posture Statement:

> These longer range (at least 5 years) projections of enemy capabilities must necessarily be highly uncertain, particularly since they deal with a period beyond the production and deployment leadtimes of enemy weapon systems. We are estimating capabilities and attempting to anticipate production and deployment decisions which our opponents, themselves, may not as yet have made.[4]

The problem is further compounded because the early years of R & D on a weapon system may produce nothing visible to intelligence gatherers. To counter this ignorance American planners tend to work on the assumption that the opposition is developing roughly the same weapon systems as themselves and also engage in very broad based exploratory research. In the words of Dr John Foster, speaking as Director of Defense Research and Engineering:

> Where threat information is adequate, we invest in amounts sufficient to meet the threat. Where information is inadequate and uncertainty high, we run some risks of over-investment to insure that our capability will be adequate, that it is sure to fulfill our strategic objectives . . . to avoid technological surprise, we must carry out vigorous, broadly-based research and exploratory development. We attempt to discover new ideas potentially relevant to national security. We test the feasibility of ideas to anticipate the worst that potential enemies could bring against us.[5]

Enough has been said to show that decision making about strategic weapon systems in general, and MIRV in particular, have to be made under ignorance. Before we can proceed to examine the MIRV decision there is just one other point which must be stressed. An actor in any decision-making process, be it an individual or a group, is motivated to seek one decision against another. This might be enough to determine the decision if the actor was the only one involved in the decision. In most decisions under ignorance, however, and especially in MIRV, there are many actors, some with opposing motivation. In order to get what it wants, therefore, an actor must *argue* with other actors, and try to *persuade* them to agree to the decision it is motivated to favour. Thus, in the MIRV case, for example, as Greenwood points out, the Air Force could not argue that MIRV should be fitted to its Minuteman fleet because the Air Force likes to play with the latest gadgets, or because without MIRV the expenditure, and hence the influence, of the Air Force would be diminished. These and similar considerations may have motivated the Air Force to seek MIRV, but in seeking it arguments had to be undertaken with other interested parties with quite different motives. It is these public arguments on which we will concentrate, not the motivation behind them.[6] Of particular concern will be the strategic arguments for and against MIRV.

In the early 1960s American strategic planners became very concerned that technological developments in the Soviet Union might soon upset the balance of the deterrent. Their fear was that the Soviet Union would deploy a sufficient number of accurate missiles to inflict severe damage on American landbased bombers and landbased Minuteman rockets. The remnants of this force would be able to retaliate, but the damage they would be able to inflict would be small, owing to Soviet anti-ballistic missiles (ABM) which would destroy incoming warheads well before they reached their targets. The development of accurate missiles and ABM might, in this way, make an attack on the United States an attractive option for the Soviet Union, as the limited retaliatory damage might be regarded as an acceptable price to pay for the military defeat of the United States.

These fears discounted the existence of a powerful force of missiles launched from submarines, which would be unaffected by the Soviet attack and able to retaliate with sufficient missiles to overwhelm any Soviet ABM. This is our first encounter with hedging in the story of MIRV. The planners could not be certain that the submarine launched missiles would be able to overwhelm a Soviet ABM; perhaps this would be much more effective than then thought possible. Nor could they be sure that these missiles would be available; perhaps some sudden failure would diminish or even annihilate the submarine fleet, or perhaps unexpected Soviet anti-submarine weapons would appear on the scene.

Failure of the submarine launched missiles would be disastrous if deterrence relied on them alone. To avoid such disasters American planners have always insisted that each element of their deterrent, the landbased missiles, submarine-launched missiles and bombers, should be an effective deterrent by itself. In the words of Secretary of Defense Laird:

> Although we confidently expect our Polaris/Poseidon submarine to remain highly survivable through the early to mid-1970s, we cannot preclude the possibility that the Soviet Union in the next few years may devise some weapon, technique, or tactic which could critically increase the vulnerability of those submarines . . . In any event, I believe it would be far too risky to rely upon only one of the three major elements of our strategic retaliatory forces for our deterrent.[7]

MIRV arrived on the strategic scene at exactly the right time. It appeared as a result of continuing engineering development, but soon took on great importance as an answer to American fears about the direction of Soviet strategic technology. MIRV offered the best counter to Soviet ABM, because it enabled the number of warheads in an attack to be multiplied until the ABM radar systems were saturated and the defence rendered ineffective. In addition, some re-entry vehicles could be fitted with jamming devices to further confuse the radar. At the same time the additional warheads enhanced the ability of the Air Force's Minuteman force to attack Soviet military targets (counterforce attack). By this time the Air Force had adopted the view that it should be able to limit damage to the United States should deterrence fail by having the ability to attack airfields, troop concentrations and remaining rocket silos in the Soviet Union; a task particularly suited to landbased missiles, bombers being too slow and submarine launched missiles too inaccurate. The possession of this so-called second strike capability was seen as part of the answer to the strategic fears about an unbalanced deterrent. MIRV would limit the damage that could be inflicted on the United States, making an attack less attractive to the Soviet Union.

There was yet a third role for the new technology. As we have seen, American fears at this time centred on Soviet deployment of missiles accurate and powerful enough to destroy much of the landbased strategic forces, so that their retaliation would be inadequate to overcome Soviet ABM. By placing a number of warheads on each missile, MIRV immediately multiplied the power of the rockets surviving a Soviet attack. This is a recurrent hedge in American strategic thinking. As McNamara stated in 1961:

> the United States has no alternative but to ensure that at all times and under all circumstances it has the capability to deter their use. In this age of nuclear-armed intercontinental ballistic missiles, the ability to deter rests heavily on the existence of a force which can weather a massive nuclear attack, even with little or no warning, in sufficient strength to strike a decisive counterblow. This force must be of a character which will permit its use, in event of attack, in a cool and deliberate fashion and always under the complete control of the constituted authority.[8]

The speculative nature of the fears which led to the development of MIRV must be stressed. It was appreciated that it would be many years before the new technology could be fully deployed; MIRV appeared on the drawing boards of its various inventors in late 1962 to early 1963 and deployment began in 1970. It was necessary for planners in the early 1960s to think what technology the Soviet Union might deploy eight to ten years ahead and whether this might require countering by MIRV, whose development would have to begin very soon if it was to arrive in time to restore the strategic balance. Forecasting of this sort is extremely hazardous, as we have seen, so American planners attempted to hedge against the worst conceivable Soviet developments. These were deployment of accurate and powerful landbased rockets, together with an efficient ABM system; and it was as a hedge against this twin development that MIRV was championed.

As discussed above, MIRV functioned as a hedge in three ways; by countering Soviet ABM, by limiting damage from a Soviet attack, and by enabling the landbased missile force to survive a Soviet attack. It is hardly surprising, therefore, to find that after a few critical skirmishes in its early days, no voice is raised against MIRV from any of the key actors in its development. This, not only because of the strength of the strategic case, but also because MIRV enabled many of the actors to further their own private goals, for instance McNamara's wish to limit the number of Minuteman rockets, and the Navy's wish to avoid involvement in plans for attacking Soviet military targets. Proponents of MIRV had, in fact, a very easy time, and were able to sell it to interested parties by stressing one strategic function, often suppressing the others. Thus, the public case, when it came, was in terms of a counter to Soviet ABM and damage limitation, while its counterforce role, sure to attract suspicion from the public, was carefully omitted. The Air Force, on the other hand, cared little for MIRV as a counter to Soviet ABM, and were sold the new technology principally on its counterforce role. Arms Control authorities were convinced of the need for MIRV by pointing to its ABM role, since they saw ABM as highly destabilizing. Its counterforce and damage limitation uses were suppressed, being objectionable to arms controllers.

It is now time to consider how views about the strategic need for MIRV changed in the light of information gathered between its inception, around 1963, and its deployment, beginning in 1970, starting with its role as a counter to Soviet ABM. American military planners became worried about Soviet ABM in the early 1960s. Construction of what appeared to be a small ABM site near Leningrad was observed in 1961, and Soviet weapon tests indicated an interest in high altitude nuclear devices which might be capable of destroying incoming warheads by their intense X-ray emissions. A second site, near Moscow, was observed in 1962, followed by worrying boasts about the Soviet's ability to shoot down attacking missiles. In 1963 the Tallinn line, a wide arc of radar installations from Riga to the far north-east of Archangelsk, was first seen. It is one thing to gather intelligence on enemy hardware, but quite another to discover what it is for. A popular, but by no means universal, interpretation was that the new radar line was part of the development of an extensive ABM system. The future development of the Soviet ABM could not, of course, be predicted but these were the signs that interest was being shown in it, and that it might become a powerful obstacle to an American attack. American planners therefore performed their characteristic hedge and, assuming the worst about the future of Soviet ABM, they turned to MIRV as a counter.

By 1967, however, most of the intelligence community was convinced that the Tallinn system had little or no ABM potential, and that it was really for bomber defence. By this time the Leningrad ABM site had been abandoned and the eight Moscow sites were clearly in the doldrums, with no signs of additional sites being prepared. To this should be added American knowledge of the enormous difficulties of missile defence, acquired from the Army's failure to develop a workable ABM system. MIRV by now was beginning to look unnecessary as a counter to Soviet ABM. Its proponents, however, suggested that it was still needed as a hedge against the rapid uprating of the Tallinn radar system. In describing the decision to develop MIRV to the 1969 Congressional hearings the Director of Defense Research and Engineering, John Foster stated:

> It should be clearly understood that we are not reacting to the Moscow ABM alone. We have to hedge against the installation of a Galosh or

improved ABM around a number of cities. *Also, we are still concerned about the capabilities of the Tallinn system. That system employs a large number of interceptors which could be converted to an ABM capability in addition to their anti-aircraft role.* The planned MIRV force is needed to maintain our deterrent posture in the absence of assurances that Tallinn has no ABM capabilities or is not being upgraded to have ABM capability. Thus far, since we continue to have uncertainty about the Tallinn-type system, we have planned our force conservatively.[9]

By 1972, when the deployment of American MIRV was well in hand, the SALT I agreement was signed which severely limited ABM deployment by both sides. The Protocol to the ABM Treaty signed two years later allowed each side one ABM site, and shortly after both dismantled their one site This had nothing to do with MIRV; it was simply that by this time missile defence was known to be unworkable.

Having considered the tangled history of MIRV as a counter to Soviet ABM we may now turn to the no less complicated story of its use as a counterforce weapon. The idea, as we have seen, is that MIRV would enable many more warheads to be used in an attack on Soviet military targets aimed at limiting the damage the Soviet Union could inflict on the United States. Some of these targets are soft, like airfields and troop concentrations, but in 1964 the real bonus of the Mark 12 MIRV was seen as its accuracy, which would enable it to destroy semi-hard targets such as the all important missile silos in the Soviet Union which had not yet been launched against the United States. The 1964 Damage Limitation Study concluded that the best way of attacking Soviet silos was by launching accurate missiles, rather than launching two or more missiles at each target.

But by the late 1960s it became increasingly clear that the first generation MIRVs were not an effective counterforce weapon. The degree of protection of Soviet missile silos had been greatly underestimated, as had the difficulties in attacking them. Hence, by the time the first MIRVs were being deployed it was known that they could play a much more limited counterforce role than the one planned for them.

The third reason for MIRV, the enhancing of the Minuteman force's ability to survive a Soviet attack, looks equally dubious in the light of the above developments. The fear was that the Soviet Union would be able to destroy much of this force on the ground and then reduce retaliatory damage to an acceptable level by ABM. But this fear faded as Soviet ABM was seen to pose no real threat and as the difficulties in launching a counterforce attack came to be appreciated.

It was apparent, therefore, that MIRV was not needed for any of the three roles it has been designed to fulfil, and yet its deployment went ahead from 1970. Shortly before this the intense secrecy surrounding the project was relaxed and the need for the new technology began to be discussed in wider circles. Two criticisms of MIRV emerged, notably from the Arms Control and Disarmament Agency and outside arms controllers. It was argued that the Soviet Union would have to place the worst possible interpretation on American deployment of MIRV, seeing it as a counterforce weapon for use in a first strike against the Soviet Union. This it was held, would seriously unsettle the balance of deterrence, even promoting Soviet intentions to launch a first strike against the United States. As a hedge against this worst outcome, the United States should abandon MIRV.

The second line of criticism was that once MIRV was known to work, the Soviet Union would be forced to assume the worst about American forces, and work on the belief that all American missiles carried MIRV. The problem was that only on-site inspection could distinguish between a single warhead and MIRV, the external appearance of the rocket and silo being the same. But such inspection had proved impossible to negotiate. The Soviet Union would have to hedge and take all rockets as carrying MIRV, while the United States would be obliged to fulfil the Soviet assumption by MIRVing all its missiles. By parity of reasoning, when the Soviet Union possessed its own MIRV, it would be obliged to MIRV all its rockets. Thus each side would be forced to deploy an enormous number of MIRVs, and problems of on-site inspection would prevent any negotiated reduction. The only escape from this absurdity, it was argued, was to refrain from testing MIRV. By 1969 this argument was heard in Congress, though with little final effect since the first tests had already taken place in August 1968.

Final opposition to MIRV flared up in 1969-70, over the need to delay deployment so as not to pre-empt its negotiated elimination at SALT I. This was, however, very ineffective as most of those urging arms control at the time decided to concentrate on an easier victim, American ABM. In addition, the fight about MIRV at this very late date was very one sided. In Greenwood's words:

> by 1968 the programmes were so far along, so many resources had already been invested, so many other things had already been foregone in deference to MIRV, so interdependent were the expected deployment of the new missiles and the continued viability of the whole deterrent force, and so much organizational and personal prestige within the

services relied on the timely and successful completion of the programmes that delay or cancellation was considered truly intolerable by the military.[10]

With its deployment, our interest in the story of MIRV ends, and we may step back to consider the quality of the decisions about its development and what lessons the story might hold about the control of technology and the obstacles posed by competition.

Lessons from the Case Study

(a) Competition made MIRV's development totally inflexible

The question to be considered now is the degree of rationality of the decisions involved in the MIRV story. These were all decisions under ignorance and so on the theory of Chapter 2 they should have been flexible. The decision makers should have been able to collect facts showing their original decisions to be wrong and should have retained the freedom to use this information in correcting their decisions. In other words, they should have been in control of the development of MIRV. This requirement seems to be totally absent from the history of MIRV, which continued to be developed and deployed after it was known that it did not fulfil any of its original functions. It is this irrationality which we must first seek to explain.

There is something profoundly wrong with decision making about strategic weapons of which MIRV is not an untypical example. The present situation in which both the United States and the Soviet Union are loaded with enough nuclear dynamite to destroy the world many times over cannot be the product of rational decision making. Somewhere, something is drastically wrong. Many commentators point to the fault as lying in 'technological determinism'. They see nuclear weapon technologies as having a life of their own, uncontrollable by, and often quite contrary to the true interests of, their creators. A typical cry of this sort is from Frank Barnaby, Director of the Stockholm International Peace Research Institute: 'The dilemma of the nuclear age is that, despite the desire of political leaderships to avoid such a war, we are being driven towards it by uncontrolled military technology'.[11]

MIRV has been seen as an example of this kind of determinism. Thus, for Tammen, 'a sound argument can be documented that technological determinism did in fact play a central role', and 'MIRV was inevitable from the point of view of being a natural accumulation of technical knowledge',[12] and York expresses the same opinion when he says that for MIRV 'Almost all the important decisions were technologically determined.'[13]

The thesis of technological determinism perhaps explains the crazy momentum of the strategic arms race, but, as expressed, it is most unsatisfactory. For a start, what is supposed to give technology such an unstoppable inertia? Decisions are made by human beings, and if technological developments constrain their decisions to such an extent as to be beyond control, what is it about this interaction between human and technology which makes this happen? Secondly, the thesis confuses origin with cause. What Tammen and York have done is trace the *origin* of MIRV in a whole network of technological developments, from which they conclude that the *cause* of MIRV is somehow not to be found in human agency but in this network of technology. In this they commit what is known as the genetic fallacy. MIRV could not have existed without the network identified by Tammen and York, but human action is needed to draw MIRV from the network, and in the absence of this action there can be no MIRV. Thirdly, the thesis only looks plausible when applied with hindsight. There are many strategic weapons systems, even some major ones, which have been partially developed and never deployed, like the American ABM system which figured in the final stages of the MIRV story. If technological determinism holds, this just should not happen. A weaker form of determinism might hold, it is true, but the least we would demand of it is that it be able to predict what bits of technology will be unstoppable and what bits controllable; something not yet, if ever, possible.

Proponents of technological determinism see the problem with decisions like MIRV as being with the irreversibility of the technology – once it has proceeded so far, it acquires a momentum which cannot be resisted. Care must be exercised here about just when the momentum becomes irresistible. Any large project is difficult or even impossible to stop at some point near its completion, because everything has been adjusted to fit it. We have seen that MIRV was in such a position around 1968. If technological determinism is to have any substance it must say more than this commonplace. Presumably it claims that the decision to develop the technology is irreversible more or less from the beginning. This was not the case with MIRV. MIRV *could* have been cancelled before its deployment. If the arguments by arms controllers in 1968-70 had centred on MIRV instead of American ABM, then MIRV might never have been deployed.

This suggests that if we see irrationality in the MIRV decision we should look, not simply to irreversibility but to monitoring as well. It is my claim that when the decision to develop MIRV was reversible, no information indicating the need for a reverse was obtainable, and that by the time this information was obtainable, the ability to reverse the decision had been lost. For this reason, the MIRV decision is totally inflexible.

Consider the problem posed by the Soviet Tallinn line. If the American strategic planners could have known with some certainty that this could not be transformed into an ABM system, then it could have been argued that MIRV could be safely abandoned. But such certainty could not be achieved before MIRV's deployment date. True, the growing consensus among the intelligence community was that Tallinn could not be upgraded, but it requires great confidence to countenance the abandonment of MIRV, far more than justified by the inherently uncertain business of intelligence gathering. The same point could be made for the counterforce role assigned to MIRV and the growing realization that the first generation MIRV could not fulfil it. By the time this was known with sufficient sureness to point to the need to abandon MIRV, MIRV was already deployed. Again, the picture painted by opponents of MIRV; that it would merely drive both sides to deploy it as extensively as possible, with the impossibility of negotiating controls on its deployment, may have been plausible, but could it have been known to be accurate with sufficient certainty for MIRV to be abandoned? Once more, whether or not this would come to pass could only be known with enough certainty to call for action once MIRV had been deployed.

Until MIRV was deployed the decision to develop this technology was reversible. Like any other major project it became harder and harder to reverse as time went on, because more and more people acted in the expectation of its arrival, but this is a feature of any technological change. Until deployment, the technology was not inevitable. But information of sufficient quality to point to the technology's abandonment was not available. This information eventually came, but only after deployment, by which time it was too late to act on it. This is where MIRV's reversibility differs from that of ordinary large scale projects. Because of the competitive nature of MIRV, once it had been deployed the Soviet Union would have to respond and any response would entail the need for the United States to retain MIRV. Thus, the Soviet Union could perceive American MIRV as a first strike weapon, whereupon they might act by increasing the number of their own missiles, but MIRVing their missiles or by protecting their silos by ABM. In each case the response by the Soviet Union exacerbates the problems American MIRV was designed to overcome. Thus, if an American MIRV was called for originally, the call for it is now even stronger. This is, of course, exactly what happened, and American MIRV led rapidly to Soviet MIRV. The situation is even more ironic because the problem of on-site inspection means that there can be no negotiated limits to the MIRVing of missiles. True, the SALT II agreement places 'limits' on MIRVd missiles, but in all cases these are so high as to have no real significance.

It follows that MIRV was never under the control of its creators. In one sense it was, because it was possible for a few years at least, to abandon the project, but this cannot be seen as amounting to genuine control. Genuine control calls for the decision makers to be able to revise their decisions in the light of the information which they receive. When MIRV was reversible, there could be no such information. In this period whether the project went ahead or was abandoned could not be decided on rational grounds – in response to information showing that the original decision did not, after all, further the objective of avoiding war. Those favouring the technology painted pictures of the future which called for the project to go ahead, such as the sudden upgrading of the Tallinn line. Those opposed to it painted other pictures, such as Soviet reaction leading to Soviet MIRV, which called for the abandonment of the project. But all this is so much fantasy. These are pictures and pictures only, and cannot offer serious guidance as to the future of MIRV. This can only come with the acquisition of firmer information, by which time, of course, the project cannot be abandoned.

At the root of this problem is the lead time of MIRV together with its competitive setting. The pattern can be summarized as below:

(i) Competition means that error costs are very great; in the case of MIRV they might include a Soviet victory in a nuclear war. For this reason it is necessary for the United States to avoid the worst that might happen in the future, in this case that the Soviet Union will be able to launch a first strike against American bombers and rockets, limiting retaliation by ABM.

(ii) Because of the lead time necessary to acquire new technology, the United States cannot delay starting on a technological answer to the worst possible future until it actually happens. Work must begin on MIRV before the feared situation materializes.

(iii) The Soviet Union eventually becomes aware of American interest in MIRV and must act in a way symmetrical with the above. The worst possible deployment and strategic use of American MIRV must be assumed, and work begun on counter technologies long before the unsettling American technology actually appears.

(iv) The Soviet response includes its own MIRV programme, so both sides end up with MIRV.

The pattern is very clear in the case of MIRV, but it underlies very many decisions about the control of less spectacular technology, where the competition is economic and not military. Competition increases the cost of error, so forcing each competitor to hedge against what his opponent might do. Where hedging requires the development of new technology with a significant lead time, each competitor finds himself hedging against a future contingency whose chief attraction lies in its provision of a hedge for his opponent against his own actions. In this way both finish with technologies which may have no intrinsic attraction at all.

Table 4.1 – Payoff to USA in Game Against USSR

USA \ USSR	MIRV	No MIRV
MIRV	xx	xxx
No MIRV	x	xxxx

Decisions like MIRV are customarily seen in terms of Bayesian game theory. The two players are the United States and the USSR, and each has two options, MIRV or No MIRV. The payoff to a player is a function of his choice and the choice of his opponent, both players making their decision at the same time and so in ignorance of the other's decision. The situation can be represented by the matrix in Table 4.1, where payoff to the United States is indicated by crosses, the absolute values being irrelevant. The matrix shows the best outcome for the United States to be neither side having MIRV, and the worst to be the United States not having MIRV and the USSR having it. The payoff to the USSR is symmetrical.

How should the United States choose? If No MIRV is chosen, the payoff may be best, if the other side chooses the same, or the worst, if the other side chooses MIRV. To avoid the worst outcome, MIRV should be chosen. The same reasoning applies to the USSR, so the final outcome is that both sides choose MIRV, giving them less than the best payoff that they could obtain. This is the so-called prisoners' dilemma explanation of decisions like MIRV; neither side wants the thing, but must avoid being without it if the other side has it.

If this interpretation of the MIRV decision and other arms-race decisions is correct, we have the singularly depressing result that the arms race is a rational enterprise. Luckily, however, the analysis does not hold water for very long. The key error is trying to see each player as making his move in ignorance of the move made by his rival. In real life, such as the MIRV story, this does not happen. Each side has time, and more importantly can buy extra time, in which to note the behaviour of his opponent and make a suitable response. Decisions by both sides can be made flexible, so that they can be reversed in the light of the decision of the other side, something alien to the prisoners' dilemma analysis. Finally, the two sides can communicate, and attempt to break down the barriers of distrust and suspicion which generate the problem they both seek to solve, something not allowed in the Bayesian analysis.

(b) Hedging is no way to cope with ignorance

The role of hedging as a way of coping with ignorance is clear in the MIRV story. All the arguments for MIRV, for example, depended on assuming the worst about Soviet technological developments and intentions. In the words of Dr Foster, the Director, Defense Research and Engineering, who was quoted earlier:

> Our current effort to get a MIRV capability on our missiles is not reacting to a Soviet capability so much as it is moving ahead again to make sure that, whatever they do of the possible things that we imagine they might do, we will be prepared.[14]

In making decisions under such extensive ignorance, hedging is enormously attractive, because it means that at least the worst which could happen has been avoided, and so it is hardly surprising to find that it occupies a dominant position in the logic of defence planning in general, and in the MIRV story in particular. Nevertheless the MIRV story reveals the ultimate absurdity of coping with ignorance by hedging. Firstly, hedging leads to contradictory prescriptions about MIRV. Hedging based on the assumptions made by MIRV's proponents, about Soviet first strike potential and ABM, calls for the introduction of MIRV. MIRV's opponents, however, suggest that the new technology could lead to nuclear war because of the destabilizing perception of MIRV by the Soviet Union as an American first strike weapon. Hedging against this obviously calls for the abandonment of MIRV. The lesson, of course, is exactly as before. Without facts to guide the decision makers, fantasies of all kinds fill the void, and action is actually recommended on the basis of fantasy. Proponents of MIRV pick their fantasy and cling to MIRV as a hedge against it; opponents design their own fantasy and recommended MIRV's abandonment.

But even if hedging was not contradictory in this way, the MIRV story points to a second failing. Hedging here led to the adoption of MIRV, but none of the developments imagined actually occurred. Despite this, MIRV continued for the reasons we have seen. If hedging is unavoidable, it can only make sense where it is possible to revise the hedge if the envisaged worst case does not materialize.

In the whole MIRV story the realization of the need for genuine flexibility is recognized only once, and then by such a minor actor as to have no influence on the decision. Martin McGuire was in the Office of Systems Analysis in 1965 when decisions were being made to continue the development of the two MIRV systems and to include them in estimates of future force. McGuire observed that the lead time for the deployment of MIRV was certain to be considerably shorter than the time needed for the Soviet Union to develop and deploy an extensive ABM system. He also feared that a hasty deployment of MIRV would merely stimulate a Soviet response which might destabilize the deterrent. He argued, therefore, that the United States could safely delay its MIRV programme until there was clear evidence of a developing Soviet ABM system, and perhaps use the threat of deploying MIRV to induce the Soviet Union into serious arms control talks.

Nothing came of McGuire's argument, and it is not possible for me to judge its merits in the light of the technical problems facing MIRV and ABM. What can be said, however, is that the argument illustrates the approach to handling ignorance which is essential if decisions like MIRV are ever to be taken in a better, more rational way. If McGuire's approach had been practicable and eventually adopted, then the United States could have safely kept MIRV in readiness for any Soviet ABM which might appear If none appeared, then MIRV need never see the light of day. In this way some sort of genuine control could be exercised over the technology. It might still be the case that once deployed, MIRV was impossible to scrap, but McGuire's approach would at least ensure that MIRV is only adopted if its needed.

Herbert York (1976) has performed a similar analysis on the American decision to build the first fusion weapon, the Superbomb first tested in 1950. The development of the bomb was suggested as a hedge against the Soviet Union developing it first. York argues that there was no need to hedge in this way, because the United States could delay development until the first Soviet test of the new weapon, after which the United States, with its greater technological resources, could swiftly catch up, with no fear of the deterrent breaking down. The situation is exactly analogous to MIRV. If the United States has a superbomb, the Soviet Union must have it, and vice versa, so the first decision to adopt the weapon is effectively irreversible. York's suggestion, however, would mean that the United States would acquire the new weapon only when it needed to, so that, to this extent at least, the weapon was under control and decision makers had some flexibility.

(c) Agreement to proceed with a project may mask disagreement about when it should be adandoned

The final feature of the MIRV decision which may illuminate some general problem concerning the control of technology is that MIRV fulfilled so many functions that agreement about its development was very straightforward. On Bayesian theory this is perfectly correct, because different actors can favour the same option for different reasons. This is all there is to the matter, and we should just adopt the favoured option and be grateful for a little agreement in a world riddled with dissent. On the perspective of the theory of decision making expounded in this work, however, things are not so simple. A decision under ignorance must not be seen as a point affair, but as extended over the time in which information which might lead to the decision's revision may be acquired; and the ability and willingess to respond to this information are essential to good decision making. In the case of MIRV, the agreement to develop the technology is facile because it masks disagreement about an essential part of the decision; under what conditions is MIRV to be abandoned? Those actors seeking a counter to Soviet ABM might agree on the need for MIRV with actors who want it for counterforce or damage limitation, but they may well disagree about what would show MIRV to be no longer needed. For the first actor, MIRV would be shown to be redundant if no Soviet ABM is developed; for the second if the accuracy and power of MIRV turns out to have been exaggerated, so that it is not an effective counterforce weapon; for the third if accurate and powerful Soviet missiles capable of attacking hard American targets do not appear. Thus the happy agreement of the early stages is almost bound to be broken in response to information emerging during MIRV's development, so disagreement about conditions for abandonment is just as profound a feature of the MIRV story as the initial agreement.

The point may be made in another way. We have seen how proponents of MIRV were able to sell the idea to different actors by emphasizing the strategic function of the new technology of interest to the actor, while suppressing functions which might be objectionable. This, of course, gives the proponents of MIRV an unfair advantage over its opponents who are never permitted to see the whole strategic picture. The debate about MIRV is, therefore, unfairly slanted in favour of its defenders and some remedy for this should have been supplied. Fairness in the debate could have been achieved by insisting that the actors agree, not only on the need for MIRV, but also on the conditions which would justify abandonment of MIRV. In this way, all the actors would have to be given the whole picture and would have to be honest with one another about their intentions and interest in MIRV.

References

1 National Computing Centre (1979).
2 Braun (1978).
3 Greenwood (1975), p. 49.
4 Quoted in Greenwood (1975), p. 86.
5 Quoted in Greenwood (1975), p. 87.
6 More on this in Part 2.
7 Quoted in Lapp (1970), p. 76.
8 Quoted in Kaufmann (1964), p. 53.
9 Quoted in Tammen (1973), p. 103.
10 Greenwood (1975), p. 139.
11 Barnaby (1979a), p. 583.
12 Tammen (1973), p. 94.
13 York (1975), p. 35.
14 Quoted in Lapp (1970), p. 68.

Bibliography

F. Barnaby (1979), 'What SALT II Means', *New Scientist*, 14 June, 905-7. (1979a), 'Inevitable Conflict', *New Scientist*, 23 August, 581–3.

E. Braun (1978), 'Microelectronics and Employment', *Proceedings on Alternatives to Unemployment*, North East London Polytechnic.

T. Greenwood (1975), *Making the MIRV: A Study of Defense Decision Making*, Ballinger.

W. Kaufmann (1964), *The McNamara Strategy*, Harper and Row.

R. Lapp (1969), *The Weapon Culture*, Norton. (1970), *Arms Beyond Doubt, The Tyranny of Weapons Technology*, Cowles.

National Computing Centre (1979), *The Impact of Microprocessors on British Business*, Manchester.

R. Tammen (1973), *MIRV and the Arms Race: An Interpretation of Defense Strategy*, Praeger.

H. York (1970), *Race to Oblivion: a Participant's View of the Arms Race*, Simon and Schuster. (1976), *The Advisors: Teller, Oppenheimer and the Superbomb*, Freeman. (1975), 'The Origins of MIRV', in D. Carlton and C. Schaerf, (eds.), *The Dynamics of the Arms Race*, Croom Helm.

5. THE HEDGING CIRCLE

Some technologies show a rate of growth which is almost spectacular. The prime example must surely be the technology of nuclear deterrence, which will be considered in one of the case studies which follow. The growth in absolute numbers of weapons and in their level of sophistication is vastly in excess of what seems required to maintain a deterrent force capable of destroying the enemy if he once attacks. Each side now possesses of the order of 11,000 strategic warheads and bombs of various sizes, enough to destroy his opponents many times over. On the face of it we seem to have an example of a technology which has completely run away from human control. Other examples of technologies which are growing at a less inflated rate, but still faster than appears reasonable might include energy supply technologies, to be discussed in the second case study, particularly electricity supply, water supply and road transport. In this Chapter, I want to suggest a mechanisim which might be responsible for the over rapid growth of technologies of many kinds, one I shall call the *hedging circle*.

This mechanism can operate on a technology *A* whose function is to provide *X* if:

(i) Not providing *X* is expensive;
(ii) The lead time for technology *A* is long;
(iii) Of all the technologies for providing *X*, *A* alone cannot fail to provide *X*;
(iv) When provided with a certain amount of *X*, the social/technological system adjusts, becoming heavily dependent on having at least this amount of *X*.

If there is a supposed need to increase the future provision of *X* then *A* is the favoured means because it alone guarantees that the provision will be increased, all other means running the risk of failure. This means that the error cost of providing more *X* by *A* is less than the error cost of providing it in any other way. The long lead time of *A* means that forecasts of future requirements for *X* have to be made; it is too expensive to wait until the supply of *X* is obviously inadequate before increasing it. When more *X* is provided the social/technological system adjusts so that it comes to depend on receiving at least this much *X*. The forecast that more is required is, therefore, self-fulfilling. If the forecast is acted upon and more *X* provided, then this turns out to be necessary to the working of the system.

When the next increase in the provision of *X* looks to be required the same pattern occurs, only more boldly. The social/technological system has adjusted to receiving more *X* than before, so that the cost of failing to provide *X* is even greater than at first. There is, therefore, an even greater need to hedge by choosing to provide the additional *X* from technology *A*, since any other choice involves a risk of not being able to produce more *X*. When the scale of *A* is expanded to generate more *X*, the social/technological system adjusts as before, in a way which fulfils the forecast that more *X* was going to be needed. There is thus a vicious circle established; as the scale of *A* increases, the system adjusts to having the quantity of *X* which it provides so that the cost of failing to provide *X* increases, making it even more important to hedge against a failure of future supply by meeting any projected increase in demand by expanding *A*. It soon becomes too risky to attempt using some other technology for the provision of the additional amounts of *X* which are forecasted as needed in the future. Thus the only way to meet future growth in demand for *X* soon becomes *A*.

This is in contradiction to the prescriptions of the theory of decision making under ignorance. Future options of using other technologies are ruled out, so the choice of *A* is an inflexible decision. Thinking in terms of the control of the social/technological system by providing *X*, the system becomes harder to control as it becomes increasingly dependent on the technology *A*. We saw in Chapter 3 that a low variety control is difficult. If there are forty ways of providing *X* apart from *A* then any unexpected failure of *A* can be easily accommodated. But where *A* is the only provider of *X* its failure cannot be tolerated. Here is the explanation of the excessive growth of technologies. The low variety control, the long lead time of *A* and the past adjustment of the system to the provision of *X* combine to make any shortfall in supply very expensive. As a hedge against this, *A* must be provided on a very generous scale. To put it another way, the system has become increasingly difficult to control so that its performance is increasingly sensitive to errors in forecasting. If forecasts are too low, it is very expensive and takes a long time to correct them. The future provision of X must, therefore, be planned so that these errors are avoided. In other words, the scale of A must be increased as a hedge against forecast errors. But as A expands, the system adjusts to it as before, so that the precautionary planning appears to have been justified.

Table 5.1 – The Hedging Circle

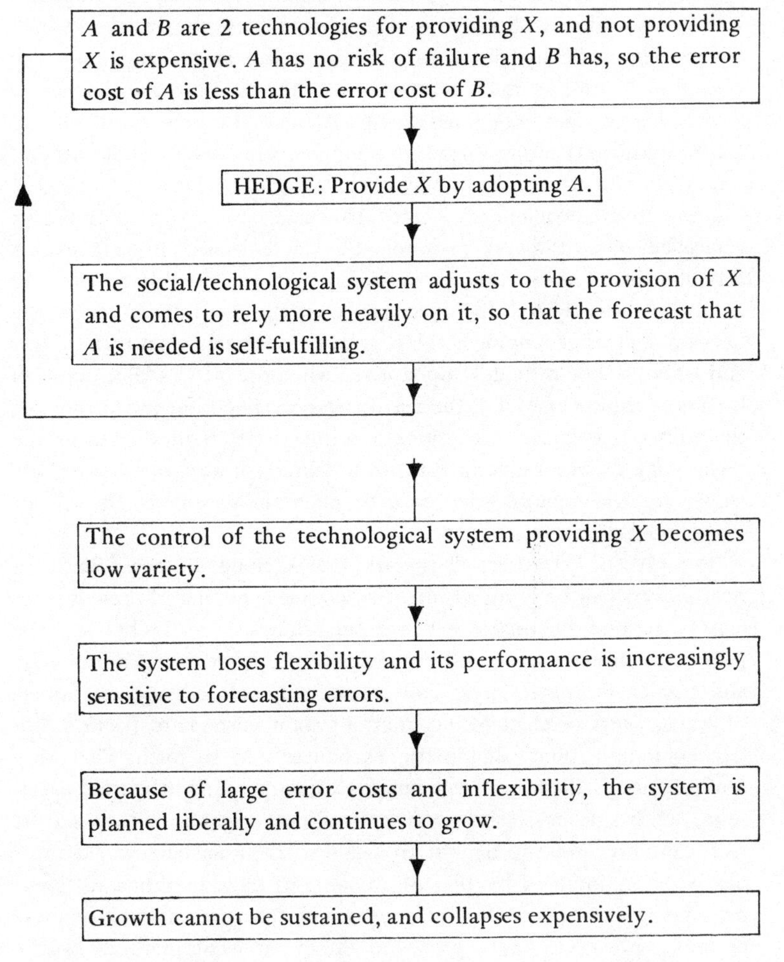

Growth like this cannot go on forever. Eventually the ability of the social/technological system to absorb ever-increasing amounts of X becomes saturated. If this happens, the system takes a long time to adjust to the new situation and a huge gap between the supply of *X* and the demand for it opens up. In worse cases the provision of *X* collapses when the raw materials used by *A* become exhausted. The whole sorry story is summarized in Table 5.1.

Case Study 1 – The Nuclear Arms Race

To the extent that MIRV is typical of decisions about strategic weapons, we can only take the direst view of the arms race. If the decisions which collectively make up the arms race are made in the appalling way typical of MIRV, we must surely ask why the arms race continues as it does. This, in fact, takes us straight back to the very centre of the MIRV decision. Fearing various developments in Soviet technology and intentions, American planners immediately looked for a *technological solution* to the problem they faced. No other kind of solution was ever canvassed. From the very beginning, the whole decision-making process was characterized by this narrowness. Whatever the differences over details, it was agreed by everyone that the solution lay with the development of technology; the only questions being what technology, and when it should be developed. Even when the MIRV–ABM question became a topic at SALT I, the narrowness remained, for the purpose of the talks was to place restrictions, and then pretty modest ones, on the technology of either side; it was not to search for ways of avoiding war which do not require each side to maintain enormous arsenals of lethal technology.

For present purposes, all the ways of keeping the peace between rival powers can be grouped into two classes. The first of these is what may be termed diplomatic methods; such things as treaties between the powers themselves and third countries, trade agreements, minor wars and border skirmishes, real and threatened trade embargoes, threats of war, cultural exchanges, negotiations about matters of friction, hot line communications, diplomatic exchanges and so forth. The other group contains all those developments which amount to mutual deterrence, each side possessing sufficient military might to ensure that its rival can never hope to benefit from a war. In an unstable world peace has to be maintained by the use of some of these methods, different ones being appropriate for different times. The use of some of these methods, however, has a profound effect on what methods can be used in the future. If the peace is threatened at some time, suppose one of the powers to opt for a policy of deterrence, acquiring the necessary weapons to ensure that no attack from its opponent can ever hope for success. The difference between weapons of deterrence and of attack lies in the intention behind them, but the other side cannot afford to trust the good intentions of the first power, even if they are real at the moment, for intentions can change like the wind. The other side must see itself as threatened and must use one of the methods of avoiding war to restore the situation.

All the diplomatic methods of avoiding war involve both powers, for example, in negotiations or in issuing and interpreting threats. As such, they all involve a risk of failure. One side may, for example, break agreements banning some new weapon and win a war with its illicit technology, or threats may be misunderstood with all sorts of complicated consequences even to war. These risks must now appear much less attractive than previously because the newly acquired arms of the other side entail that the cost of failure, that is the cost of war, is now much bigger than before. Thus the option of deterrence is favoured, because this does not suffer from the same dangers. The decision to acquire deterrent weapons involves only one power, not both, so there is no other party whose deviousness can introduce a risk of failure. Unlike the diplomatic methods, deterrence is sure to work.

Both powers now have deterrent forces, but these are not static and it will eventually happen that the balance of terror will shift to favour one side. How can the weaker side ensure the avoidance of war? It is, of course, in exactly the position of the weaker power in the above paragraph. Diplomatic methods of maintaining peace, being inherently risky, must be even less attractive than before, because the cost of failure is now even higher, the other side having a greater advantage than previously. The option of restoring deterrence is, therefore, more highly favoured than previously as it is the only way to avoid these great risks. In this way the first decision to maintain peace by deterrence biases the second decision to be made in the same way so that a vicious circle exists. The next time deterrent forces become unbalanced the cost of war is even higher because more weapons exist, and so restoring deterrence is even more favoured than at earlier times. The diplomatic options open to the powers gradually close as their failure becomes more and more expensive, until the only way of maintaining peace is through mutual deterrence.

In the previous Chapter it was observed that decisions about the acquisition of deterrent weapons are ones which have to be taken under ignorance, because huge uncertainties exist about the other side's future capabilities and intentions. As such, options should be favoured which leave future decisions open so that some response can be made as these uncertainties are resolved with the passage of time. These are options which can be quickly corrected so that the cost of error is low and the system's performance insensitive to error. Once the path of deterrence is taken it is increasingly impossible to adopt these options. Error costs rise as more and more weaponry is deployed, options of using diplomatic methods to maintain peace in the future gradually disappear because they pose greater and greater risks, and the performance of the peace-keeping system becomes increasingly sensitive to error as the cost of war increases. The path of deterrence is as rational as it is easy to get off.

The irrationality of deterrence is clearly seen in the uncontrollable growth which it always exhibits. As we have seen in Chapter 3 if there is only one way of doing something and if the costs of not doing it are large, then it must be made as certain as possible that the way will work. When weapons are the only way to keep the peace, and their very existence makes war hugely expensive, then either side must be sure beyond doubt that its weapons are a real deterrent; no possibility of failure can be countenanced. Each side is thus forced to hedge against even remote possibilities of failure; it needs to know that it can deter its opponent whatever technology the other deploys, whatever miscalculations it and its opponent make about their weapons.

Technically we may say that deterrence is a low variety control. If deterrence is once used to maintain peace, then it quickly becomes the only way to do this, so that there is no substitute if it fails. If diplomatic methods had been used throughout, the control exercised would have been of high variety because if one method had failed, others might be turned to. The need to hedge against failure would not exist because the cost of failure, i.e. of war, would not have been inflated by repeated attempts to restore deterrence. High variety control is stable and low variety unstable. The need to hedge once deterrence has been adopted leads to the indefinite growth of deterrent forces; there is always some danger on the horizon which must be countered by developing more weapons or weapons of greater sophistication because no risk is tolerable. But indefinite growth cannot be sustained. Accidental war or deliberate war following the failure to maintain the balance of terror is certain to happen some time, deterrrence cannot be maintained *forever*. A reliance on diplomatic methods of avoiding war is just as certain to fail at some time or other, but the cost of its failure is far less than that of the failure of deterrence.

Just this pattern is exemplified by the nuclear arms race. Soon after the Second World War there were many ways in which peace could have been maintained. One such way was mutual deterrence by nuclear weapons, and this option was eventually taken because neither side could risk trusting the other, trust of some kind being essential for diplomatic methods of peace keeping to be used. But once the path of deterrence was taken it became impossible to avoid war in any other way. The use of diplomatic methods requires trust, but the existence of nuclear arsenals imposes such a cost on misplaced trust that trusting the other side is simply too dangerous. War, therefore, continues to be avoided by the one option which requires no trust, deterrence. Decisions to develop nuclear weapons in the early years after the War destroyed any hope of ever being able to avoid deterrence in the future and any chance of finding other ways of keeping the peace. These early decisions destroyed future options and imposed on the world the inflexibility of the nuclear arms race.

This point was quite lost on the early advocates of a negotiated ban on nuclear weapons. The famous Franck Report of 1945, of which more in Chapter 8, stated that the effective international control of nuclear weapons:

> is a difficult problem, but we think it soluble. It requires study by statesmen and international lawyers, and we can offer only some preliminary suggestions for such a study.
>
> Given mutual trust and willingness on all sides to give up a certain part of their sovereign rights, but admitting international control of certain phases of national economy, the control could be exercised on two different levels.
>
> The first and perhaps simplest way is to ration the raw materials – primarily the uranium ores . . .
>
> An agreement on a higher level, involving more mutual trust and understanding, would be to allow unlimited production, but keep exact bookkeeping on the fate of each pound of uranium mined . . .
>
> One thing is clear: any international agreement on prevention of nuclear arms must be backed by actual and efficient controls. No paper agreement can be sufficient since neither this nor any other nation can stake its whole existence on trust in other nations' signatures. Every attempt to impede the international control agencies would have to be considered equivalent to denunciation of the agreement.[1]

This optimism is sustainable only because the authors have failed to see that the existence of even the know-how to make bombs, let alone their existence, makes trust hugely risky, far more so than in the past. Misplaced trust might have led in the past to major embarrassments or minor wars, but it could now lead to a bloodless defeat by blackmail. If one side broke trust and acquired nuclear weapons when the other side had none, even a handful would be enough to coerce the weaker power on any major point of international conflict. In modern terms, suppose that just one nuclear missile submarine escaped scrutiny after an international ban on nuclear weapons. This might carry sixteen rockets, each with fourteen warheads, each warhead of 50 kilotons explosive power – a total of 560 Hiroshimas.

As might be expected from the analysis of deterrence above, the nuclear arsenals of both sides have grown enormously as military planners have developed new weapons as hedges against what the other side might deploy. This is very clear in the MIRV story of the previous Chapter. The United States feared that the Soviet Union might soon be able to deploy accurate heavy missiles, able to destroy much of the American landbased forces in a first strike, limiting retaliatory damage by ABM. This discounts the very large American submarine missile fleet which would be able to retaliate, but American planners felt that they had to hedge against an unexpected failure of this portion of their nuclear arsenal. As fears about an existing Soviet ABM system began to recede in the light of intelligence, the need to hedge against a sudden Soviet breakthrough in this technology was urged, and further hedging was automatically applied in attributing any Soviet technological developments to the worst possible motive, and in assuming the best possible performance of Soviet weaponry, and the worst possible for that of the United States. MIRV was provided as a hedge against Soviet developments which only occurred because of MIRV. American possession of the new technology forced the Soviet Union to deploy its own MIRV, to harden its missile silos and to increase the number of its missiles as a hedge against American MIRV being used as a first strike weapon against it. Thus the armouries of both sides expanded in a way which is not reversible; MIRV cannot be abandoned even though the original fears behind its development were not realized.

The Soviet Union now has 1,398 intercontinental ballistic missiles, 608 fitted with MIRV; 950 submarine-launched ballistic missiles, 144 being MIRVd; and 156 strategic bombers. Against this is balanced an American force of 1,054 intercontinental ballistic missiles, 550 of which are equipped with MIRV; 656 submarine-launched missiles, 496 of which are MIRVd; and 576 strategic bombers. All of this is beside the thousands of tactical warheads possessed by each side, even though some of these are bigger than warheads classed as strategic. Such figures are best left to speak for themselves.[2]

Table 5.2 – The Hedging Circle in Nuclear Deterrence

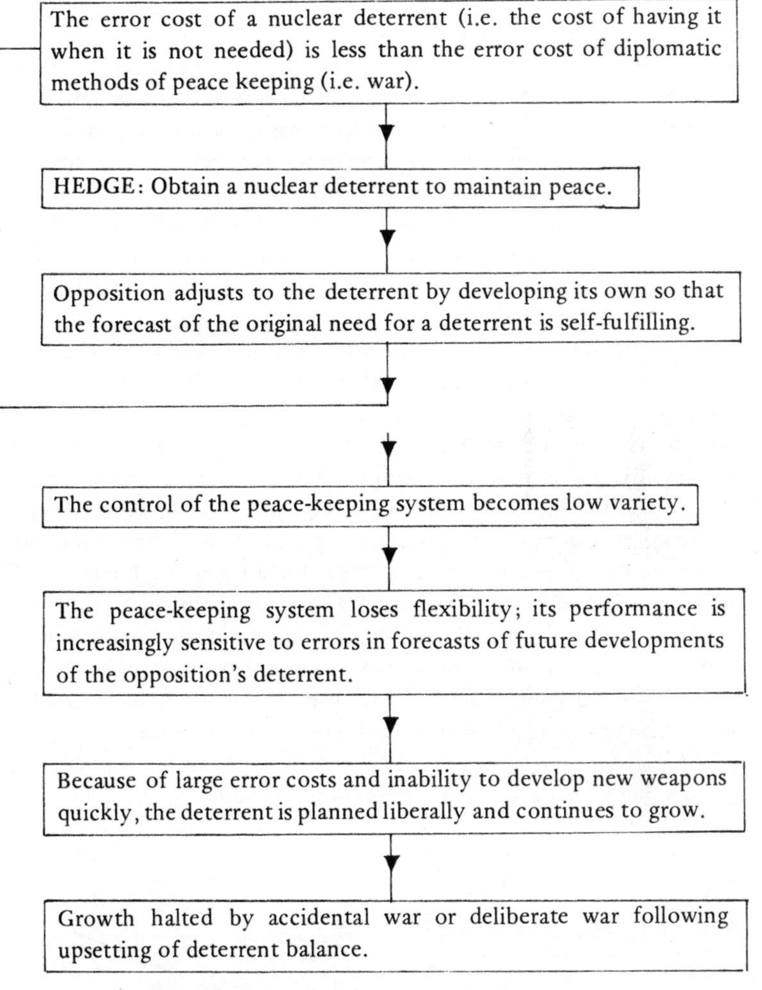

Deterrence by nuclear weapons of ever increasing size and complexity is inherently unstable. It cannot succeed *forever*, and yet it must because all options for keeping the peace in other ways were destroyed a long time ago, and a single failure of deterrence marks the end of civilization, if the term may be excused. Thus the nuclear arms race shows all the characteristics identified at the beginning of the Chapter, as can be seen from Table 5.2.

Case Study 2 – Energy Futures

There are two very different views about how the United Kingdom's energy system should develop from now until the first part of the next century. Energy is essential to all economic processes, and the aim of most energy planners is that at no time should shortages of energy restrict the country's economic growth. The conventional or high energy view is that this can only by achieved by very substantial increases in energy consumption, the chief problem being finding adequate supplies of primary fuels such as coal, oil and uranium, and building plant to convert them into suitable carriers of energy, such as nuclear power stations to make electricity. Behind this view lies a method of forecasting energy demand which tries to extrapolate past relationships between energy consumption and economic growth, making allowances for such things as increasing prices, improvements in the efficiency with which energy is used (its end use efficiency) and so on. The forecasters do not pretend that energy consumption can be forecast over fifty years or so in any great detail, but they employ scenarios which together show that energy use must grow considerably if the economy grows at even modest rates. Some typical scenarios will be considered later in Chapter 9. Details of the methods used can be found in Energy Policy (1978), Department of Energy (1978) and Starr and Field (1979). Figure 5.1 is taken from the first of these, and shows the growth in primary energy required if the British economy grows at an average of 3 per cent per annum and oil doubles in price, conservation and improvement in end use efficiency producing a 23 per cent drop in energy required at the end of the forecast period.

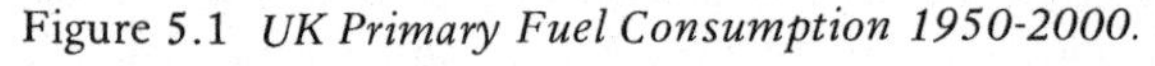
Figure 5.1 *UK Primary Fuel Consumption 1950-2000.*

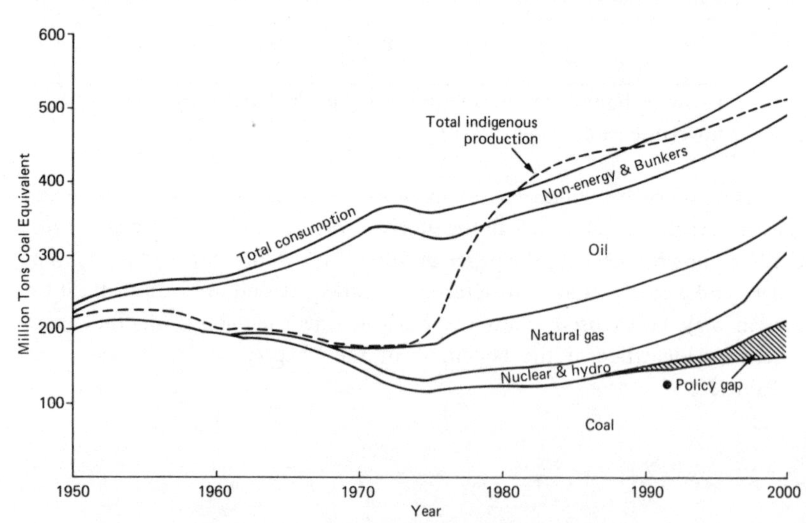

Proponents of the opposing, or low energy view are highly critical of these forecasting methods. In the words of one of them:

> Forecasting energy demand by traditional econometric techniques is now seen to lack validity. The historical linkage between energy consumption and economic output, allied to vague suppositions about the future price of fuels, is increasingly under attack. Likewise, the emphasis on supply forecasting and the policy decisions resulting from this kind of approach are palpably unsatisfactory since they are concerned only with the provision of more and more fuel, irrespective of actual consumer demand.[3]

The forecasting methods favoured by this group are far more disaggregated than those employed before. Leach explains that:

> Essentially, the approach is to start wherever possible with the ultimate purpose for which energy is used – the useful energy demand – and work upwards from there to primary energy supplies, fuel by fuel, and sub-sector by sub-sector.
>
> The forecasts start from a detailed breakdown of energy use in the baseline year 1976 ... into nearly 400 separate categories determined by end uses, fuels, and appliances.
>
> We next projected a range of basic energy-using activities. In industry this ... (was) the (net) financial output of each industry. These outputs grow in stated relationships to GDP. In transport, the activities are traffic levels for 11 types of vehicle, usually as vehicle – or tonne – kilometres. In the commercial and institutional sector the activity levels are given by the floor areas for the different building types. In housing a more complex model is used in which the dwelling stock is projected by demographic trends . . . and the activity levels are quantifiable parameters such as average internal temperatures, dwelling size and volume, quantity of hot water used per person, the amount of cooking, and the ownership of seven categories of electrical equipment. Chemical feedstocks, other non-energy uses such as bitumen and lubricating oil, agriculture, and international shipping (bunkers) are also included in the model.
>
> To turn these projections of activities into quantities of energy one has to forecast their future energy intensities – the amount of energy used per unit of activity.
>
> Given the activity level and the energy intensity, simple arithmetic gives the total amount of purchased fuel (coal, oil, gas or electricity), which we call the *delivered energy*. It then only remains to convert delivered energy to primary fuels by making assumptions about the losses in the energy supply industries and the fuel mix for electricity generation.[4]

Leach's study assumes that GDP will grow in much the same way as supposed by high energy forecasts, doubling from now until 2025 in his low case and trebling in his high case. Numerous assumptions are made about living standards:

> Within these projections of economic growth we have made numerous detailed assumptions about improved living standards. Houses become warmer so that everyone enjoys the amenities of the better-off today. Most families come to own freezers, dishwashers, clothes dryers, colour TV, and other heavy users of electricity. Car ownership grows rapidly so that 72-75% of households in 2000 have at least one car, compared with 58% today. Traffic growth in the Low case closely follows the current official forecasts for transport and road construction planning; in the High case it approaches the 'maximum possible' level. Air traffic grows 2.4-3.0 times by 2025. Total industrial output increases by a factor of 1.7-2.2. The areas of offices, schools, hospitals, shops and restaurants increase by anything from 30-80%.[5]

It is assumed that the forecasting period will see the extensive introduction of energy-saving techniques in all sectors of the economy, the technology for which either exists now or is just over the horizon. Most of these technical improvements in efficiency are now being introduced because of high energy prices, and so will become increasingly popular as prices rise in the future. Some modest government direction of this improvement in efficiency is assumed by, for example, a tightening of the Building Regulations and the laying down of energy performance regulations for cars.

Table 5.3 – Forecasts for Primary Energy Demand 2000

Fuel	Demand in 2000 AD (billion therms)	
	Energy Policy (1978) High case	Leach (1979) High case
Solid fuels	17.0	14.5
Gas	17.8	11.9
Liquid fuels	26.5	22.8
Electricity	15.8	8.0
Total	77.1	57.2

Using this approach it seems that much less energy is needed to sustain the GDP growth envisaged than traditionally thought. Table 5.3 compares total demand for primary energy as estimated for Leach's High case, and as calculated in the traditional way as in Energy Policy (1978).[6]

The consequences for energy policy are profound. For example: by 2000 the United Kingdom could be self-sufficient on North Sea oil and gas even on central estimates of reserves; coal production need only be about 120 million tonnes per year, well below the 170 million tonnes required on the basis of traditional forecasting; only about 30 GW of electricity capacity need be built until the end of the century, compared with 83 GW on traditional forecasts with a capital saving of about £30,000 million; electricity demand can be met by building only 6 GW of nuclear capacity in the first quarter of the next century so that the nuclear energy question becomes peripheral; and fast breeder reactors can be shelved indefinitely.

Our concern is not to decide which side offers the best forecasting methods, but rather to consider the problem of making decisions about the energy system in the light of the existence of such divergent views of the future. It will be best to begin with a general discussion of the difficulties of balancing demand and supply in cases where increasing supply has a significant lead time, so that forecasts of future demand must be made, and where failure to meet demand is expensive. This is typical of many technological systems including the energy system, water supply system and surface and air transport systems.

Consider a system which fulfils some purpose by the consumption of some commodity in the consumption sub-system, the commodity being provided by the supply sub-system. There will generally be very many consumers, in a rich variety of circumstances, for every supplier. The capacity of the system, the extent to which it fulfils its purpose, is expected to increase and this can be achieved in a whole spectrum of ways, the extremes of which are an increase in supply with no improvement in the efficiency of commodity use by the consumers, and better use of the commodity by consumers with no increase in their consumption. Both increasing supply and improving the efficiency of consumption have significant lead times, and failure to increase the system's capacity is very expensive.

We may suppose that forecasts of both kinds exist, one set pointing to the need to increase commodity supply, and the other set indicating that no large increase in supply is required given fairly modest changes in the habits of the commodity's consumers. There is an important asymmetry between the two changes thought by forecasters of either school to be sufficient for increasing the system's capacity. If the supply of the commodity is increased there is no risk that shortages will restrict the growth of capacity, but if the expansion of capacity is left to improvements in the efficiency with which the commodity is used there is a risk that the forecast improvements may be impossible to achieve in practice. If the forecasters prove to have been over-optimistic and efficiency cannot increase by as much as they suggest, capacity will be restricted by a shortage of the commodity until supplies can be increased, which will take some time. The cost of the forecast error is, therefore, high. This must make the risk-free option of increasing supplies attractive as a hedge against the failure to realize the expected improvements in efficiency of consumption. Suppose, therefore, that this option is chosen and commodity supplies increased to allow for the growth in capacity.

The consumers of the commodity now have no incentive to improve the efficiency with which they use it, so the forecast that an increase in supply is necessary to increase the system's capacity is self-fulfilling. If it is acted on, capacity increases by increasing the consumption of the commodity, as forecast, because there is no incentive to improve the efficiency of consumption. The consumers adjust to a plentiful and cheap supply of the commodity by substituting it for some of their other inputs, by not investing in private supplies of the commodity, by making investment decisions on the assumption that supplies will remain cheap and plentiful, and so on. The problem of how the next increase in the system's capacity can be best brought about soon arises; should supplies be increased still further or should the efficiency of consumption be improved? The need to hedge is now even more pressing than before. If it is hoped to increase capacity by improvements in the efficiency of consumption and if these prove to be impossible to achieve, the cost of the error is even bigger than before because of the various adjustments consumers have made to having cheap and plentiful supplies of the commodity. In addition, much may have been learned about the best ways to increase supplies, but nothing at all has been discovered about how the consumers can improve their efficiency. The risk-free option of a further increase in supplies is, therefore, highly favoured.

There is obviously a vicious circle here. Supplies are increased, the consumers adjust to having plentiful supplies at low prices, it becomes even riskier to reply upon improvements in their efficiency to expand the capacity of the system, so the next expansion is catered for by increasing supplies, and back to the beginning. In this way, increasing supplies at one time biases all future choices in the same direction, the bias increasing as time goes on, so that eventually the only way to increase the system's capacity is to increase supplies of the commodity, all other options being closed. There is thus a single control on the system's capacity and worse, one of low variety as there are only few ways to increase commodity supplies. Control through improving the efficiency with which the commodity is consumed is of much greater variety because there are a large number of consumers in a wide variety of circumstances. If one group of consumers has unexpected difficulty in improving efficiency, some other group is sure to find it unexpectedly simple, so control in this way should be easier than control through increasing supplies, and yet it has become impossible to operate in this way.

Where low variety controls have to be used, the control tends to grow much more than is necessary. In the present case, supplies of the commodity required in the future will have to be forecast with a liberal margin of error. If supplies are inadequate, the consumers will not be able to adjust quickly to the unexpected drop in supply, and it will take a long time to increase supplies, so the cost of error is very high. As a hedge, future supplies will tend to be calculated very generously. Supplies will, therefore, grow, but not indefinitely for this is impossible. At the best, the ability of the consumers to absorb the ever increasing supply will be saturated, with the opening up of a large and expensive gap between supply and demand; at worst the system will grind to a halt through exhaustion of raw materials.

The problems of forecasting future demand and the improvements in efficiency of consumption which are likely make decisions about the system's future ones under ignorance, so the rationality of the above model can be assessed on the theory of Chapter 2. The negative result should come as no surprise. In making such decisions options should be favoured which make the system in question easy to control, which leave future options open, and which are easily corrected so that the system's performance is not sensitive to error, all these requirements being equivalent. We have seen, however, that taking the path of increasing commodity supplies leads to a system which is difficult to control, because increasing supplies is a low variety control. The higher variety control of improving the efficiency of use of the commodity is, as we have seen, eventually ruled out. Future options are closed as it becomes increasingly risky to rely on improvements in the efficiency of consumption for the expansion of the system's capacity, and the only way of achieving this becomes increasing supply. As consumers adjust to the ever increasing supply of the commodity at low prices the cost of failing to meet their demand, i.e. the system's error cost, increases so that errors become more difficult to correct. In the same way the performance of the system becomes increasingly sensitive to errors in the forecast of future demand. The system becomes less and less controllable, and its performance ever more sensitive to forecasting errors, and decisions about it become less and less flexible and less and less corrigible.

We may now return to the debate about the future of the British energy system. This is part of the economic system whose capacity is measured by GDP. GDP can grow if energy supplies are increased, end use efficiency remaining constant, or if efficiency increases with no extra supply, these being the extremes of the spectrum of possibilities. Proponents of a low energy future favour the latter, the orthodox view seeing a need to increase supplies. The traditional view has dominated energy planning in the past, and the voice of supporters of the low energy view has only recently been heard. The system has, therefore, developed largely in the way outlined above, with the emphasis on forecasting demand by the detection of stable relationships between demand and GDP, what little concern there has been to promote end use efficiency being in response to the 1973 energy crisis. It is not surprising therefore to find that the case against the low energy option follows exactly the reasoning in the above model. The principal objection is that we cannot be sure that consumers of energy will improve their efficiency and that if the forecasts of the low energy supporters are wrong in this it will be hugely expensive in fore-gone economic growth. If supply proves to be inadequate, growth in GDP will be restricted for the considerable time needed to secure additional supplies. Because we cannot be certain about what improvements in efficiency can be achieved, this is a risk that is not acceptable. The no-risk option of increasing energy supply should, therefore, be favoured.

The case is succinctly made by Mr Norman Lamont, Parliamentary Under-secretary in the Department of Energy defending the Government's programme to build more nuclear power stations against Mr Walter Patterson, who has just put the case for a low energy future, in the television programme *The Nuclear Power Debate* (BBC1, 8 January 1980). He states:

> I am very pleased that Mr Patterson says that he is not postulating total certainty, but with great respect that seems precisely what he must be advocating [sic] . . . because what is being suggested from the other side is that the Government's modest programme for nuclear power is too much and therefore I must assume that he wants no nuclear power. And if we are to have no nuclear power I would like to know how it is Mr Patterson assumes that we will be able to satisfy our energy needs if, for example, his view of the future turns out to be wrong. What happens if the arguments which he's put forward just don't work out; how are we going to provide the energy needs of this country? . . . Walter Patterson has put forward the view that the answer

to our energy gap in the future lies in conversation, but how can he be sure that people are going to behave as he assumes they will behave? How can he be sure that people aren't just going to turn up the temperature in their houses even when they're consuming [more efficiently]? How can he be sure that even though large refrigerators consume less energy that people don't prefer small refrigerators? How is he going to be sure that people actually like to have heat pumps in their living rooms? How can he be sure about these things?[7]

Table 5.4 – The Hedging Circle in Energy Planning

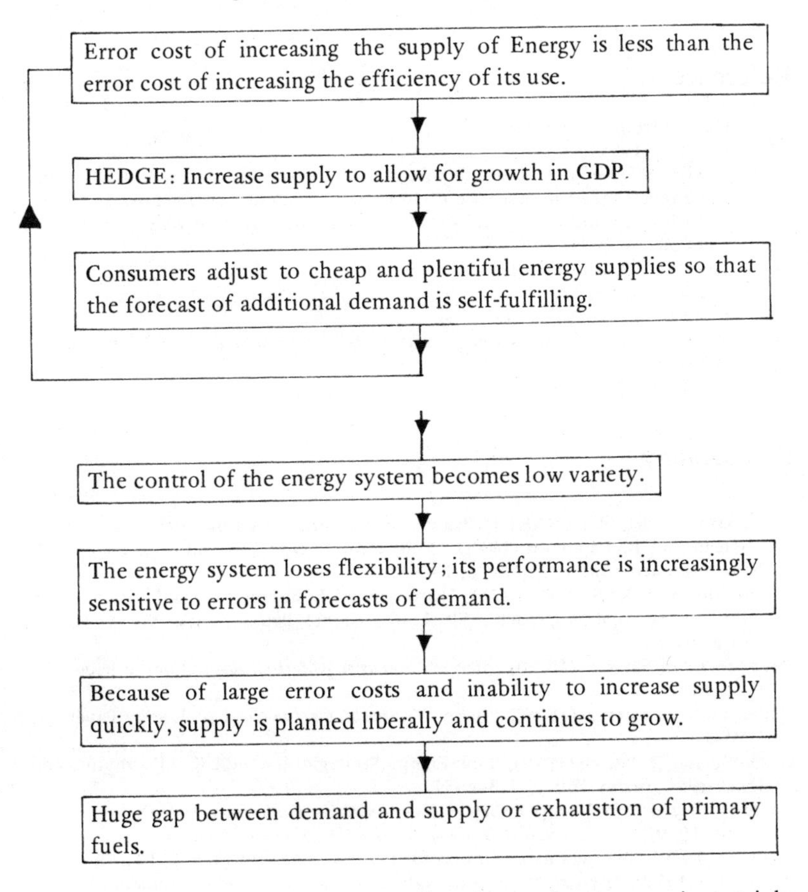

Here is the dilemma in energy policy. Lamont's argument is certainly a forceful one. Given the historical development of the energy system, much has been discovered about the best ways to increase supply but nothing learned about the ease with which the efficiency of energy use can be improved, and the cost of error in the forecasts favoured by Patterson will be very large. On the other hand, Lamont could reappear with his argument every time an increase in GDP is being discussed, forcing the energy system to become more and more difficult to manage and more and more sensitive to errors in forecasts of demand until a huge gap opens between supply and demand, as might have already happened for electricity,[8] or, worse, easily available primary fuels become exhausted. I hope to point in the general direction of a solution to this dilemma in Chapter 9. Table 5.4 summarizes the problem.

References

1 Quoted from Smith (1965).
2 Barnaby (1979).
3 Lewis (1979) p. 131. Lewis' report is an abbreviated version of Leach (1979). For other similar analyses see Chateau and Lapillone (1978), Lovins (1979), and Cheshire and Surrey (1978). For a summary of their views see Department of Energy (1979).
4 Leach (1979) pp. 11–12.
5 Leach (1979) p. 13.
6 From Leach (1979) p. 239.
7 For a quantitative approach to part of this argument see Doyle and Pearce (1979).
8 See Chaper 7.

Bibliography

F. Barnaby (1979), 'What SALT II Means', *New Scientist*, 14 June, 905–7.
B. Chateau and B. Lapillone (1978), 'Long-term Energy Demand Forecasting – A New Approach', *Energy Policy*, 6, 140–65.
J. Cheshire and A. Surrey (1978), 'Estimating UK Energy Demand for the Year 2000: A Sectoral Approach', SPRU Occasional Paper Series, No. 5, Sussex University.
Department of Energy (1978), *Energy Forecasting Methodology*, Energy Paper 29, HMSO.
Department of Energy (1979), *Energy Technologies for the UK*, Energy Paper 39, HMSO.
G. Doyle and D. Pearce (1979), 'Low Energy Strategies for the UK – Economics and Incentives', *Energy Policy*, 7, 346–51.
Energy Policy (1978), *Energy Policy – A Consultative Document*, HMSO.
G. Leach (1979), *A Low Energy Strategy for the UK*, Science Reviews.
C. Lewis (1979), 'A Low Energy Option for the UK', *Energy Policy*, 7, 131-48.
A. Lovins (1979), 'Re-Examining the Nature of the ECE Energy Problem', *Energy Policy*, 7, 179–98.
A. Smith (1965), *A Peril and a Hope*, University of Chicago Press, 560–72.
C. Starr and S. Field (1979), 'Economic Growth, Employment and Energy', *Energy Policy*, 7, 2–22.

6. LEAD TIME

In Chapter 4 the disastrous consequences of competition involving technologies with long lead times were explored through the case study of MIRV. But lead times can present problems in controlling a technology even in the absence of competition. To put the point simply; where mistakes are unavoidable it pays to discover them early and to correct them quickly. Formally; in decision making under ignorance options with a low monitor's response time and low corrective response time should be favoured. This provides good reason for the close scrutiny of sophisticated technological projects with long lead times. A striking contemporary example is the British programme of Advanced Gas-cooled Reactors (AGR). These were planned in 1964, and five were eventually started. At the time or writing two are finished, but have operated only for a short time and with many potentially serious problems.

The long lead time of the reactors has created many difficulties. They were built on the expectation of cheap uranium and expensive coal, but sixteen years after the first plans nobody can say with any assurance what the cost of the electricity is, or even whether it is lower than that produced from coal.[1] The calculation of cost requires several years' experience of operating AGRs, as many teething troubles occur in the first two or three years, and so cannot yet be done. This produces three distinct difficulties. If the cost is eventually found to be so high as to call for the abandonment of the programme, this is learnt so late that the whole energy system has adjusted to the expected arrival of AGR electricity, for example, by not building coal-powered stations, and it must now readjust. The longer the learning process, the greater the trauma this involves. It is perhaps, lucky that the AGR programme developed during a time of an increasingly large surplus of electricity producing plant.

The second problem is that knowledge of costs is essential in planning the next group of generating plants, whether these are to be AGRs or not.

The longer this knowledge is delayed the more difficult it becomes to organize the future of the system. The third problem goes back to the planning of the AGRs. Their need depended on projections of the cost of uranium and coal over the lifetime of the programme. Assuming a thirty-year life for the reactors, its lead time meant that these projections were required over close to a fifty-year period. Forecasts of this sort are, of course, notoriously difficult, but here the problems were compounded by the length of the lead time.

The case study which follows considers the problems which lead time produces for another piece of nuclear hardware, the breeder reactor.

Case Study – The Breeder Reactor

There are two kinds of nuclear reactor available for the production of electricity, thermal and breeder reactors. In the early days of nuclear development breeder reactors were thought to offer a better future than thermal ones and the first reactor to produce electricity was of the breeder variety. Military requirements for plutonium in weapons and unforeseen engineering problems, however, soon reversed this judgement, and all commercial electricity producing reactors installed throughout the world are thermal ones. In Britain breeder developments amount to the small Dounreay Fast Reactor and the Prototype Fast Reactor, producing 250 MW of electricity from 1975. In other countries there are five programmes involving similar prototype reactors in operation, in development or in mothballs.[2] Despite this humble position, breeder reactors are seen by many as essential large scale components of the future energy systems of all developed countries.

The reason for this lies in the different efficiencies with which thermal and breeder reactors consume their uranium fuel. A breeder reactor can obtain 50–100 times the energy which a thermal reactor can obtain from a given quantity of uranium. Against this, breeder reactors are more expensive to build than thermal ones. It is, therefore, necessary to trade off this extra expense with the savings from the breeders' efficient use of fuel. The world envisaged for large breeder programmes is one where uranium reserves have been seriously depleted by thermal reactors already built or under construction, so that it is an expensive commodity. It should be noted that known uranium reserves are just about adequate to fuel all the world's present thermal reactors throughout their lives. It is possible that large high grade reserves will be found in the extensive areas of the world so far unprospected, but there is no guarantee of this. It is also thought that by the time large breeder programmes are possible, oil reserves will have been seriously depleted so that oil cannot substitute for expensive uranium. Difficulties are also foreseen in any attempt to increase coal production sufficiently to substitute for uranium, so that the only way to ensure adequate supplies of energy is through the efficient use of uranium which results from its consumption in breeder reactors. A second advantage of breeder reactors is that they produce electricity directly, and demand for this extremely versatile form of energy is expected to increase greatly from now until the early years of the next century. The conventional view is well put in a Department of Energy publication thus:

> the trend of demand for electricity and doubts about the availability of fossil fuels (and other alternatives) on a sufficient scale point to the need for a large and increasing nuclear component in our energy supplies by the turn of the century.
>
> . . . the United Kingdom has no important uranium resources of her own, and if it remained totally dependent on thermal reactors it would become increasingly vulnerable to the world price and availability of uranium. Fast reactors would reduce the impact of increases in uranium prices and reduce the possibility of our not being able to supply our demand for uranium from the world market. It seems essential, therefore, to keep open the option of using them.[3]

Not only would breeders use existing supplies of uranium very efficiently, they would also produce energy so cheaply that it would become worthwhile to exploit grades of ore too poor to utilize at the moment.[4]

These projections of the necessity for breeder reactors on a very large scale have not gone uncriticized. Doubts have been expressed about the reality of the energy gap the breeder is supposed to fill, about the assumed scarcity of uranium, the safety of breeder reactors, the problems of siting a large number of potentially dangerous reactors in a heavily populated country like the United Kingdom, health hazards from the transport and fabrication of large quantities of plutonium and from waste disposal, and the ease with which nuclear weapons could be made using plutonium from breeder reactors. It is not my intention to enter a debate where I am not qualified to speak. I merely want to consider one question about the breeder reactor, namely what elements should be included in the cost of the reactor?

To answer this question the real world with all its horrible complications and uncertainties does not have to be considered. This is necessary for a numerical estimate of cost, but the present intention is just to see what things should be included in this calculation, so a simplified world is all that is required.

In this simplified world, let there be three primary fuels available to the United Kingdom, coal, oil and uranium. Since North Sea oil has been used up, all the country's oil is imported, and as reserves shrink prices are expected to rise dramatically. It is therefore planned to substitute uranium for oil, but the above-mentioned uncertainties about the cost and availability of uranium have led to the decision to burn the uranium in breeder reactors. These will be able to use the country's huge stock of depleted uranium and so make it independent of foreign supply. Three things limit the rate and scale of this breeder reactor programme. The scaling up from 250 MW(e) to, say, 1,300 MW(e) involves many engineering problems, so that it will be prudent to test designs by building a first reactor, not commencing further ones until the first is completed. It may be supposed that the first reactor takes fifteen years to commission, later ones taking ten years. The second restriction is from the capacity of the nuclear construction industry, and the third from supplies of plutonium. Existing stocks of plutonium in the United Kingdom are probably enough to begin fuelling about five 1,300 MW(e) reactors,[5] after which further plutonium must be supplied from unlikely imports, from destruction of nuclear bombs or from breeder reactors already operating. It is probably generous, but suppose that a start is made on a new reactor every second year.[6]

What we should have learned to ask by now is the question of how long it is before the decision to embark on this programme can be detected to have been in error? There are, of course, very many kinds and degrees of error; the reactors may simply be a little more expensive to build than at first thought, or they may pose slightly higher than expected risks to their operators. For present purposes, consider the grossest kind of failure; the reactors and/or the reprocessing cycle they depend upon simply does not work. In detail, suppose the first reactor fails to produce electricity, but that this is regarded as due to a fault which can be eliminated in the second reactor. Design modifications are made but the second reactor also fails to function. This, however, is only learned when it has been completed, twenty-five years from the beginning of the programme. By this time work is underway on reactors 3, 4 and 5. Desperate attempts are made to avoid the earlier failures, but this proves difficult as work has already begun. It continues until reactors 3, 4 and 5 are finished at years 27, 29 and 31, but none is able to produce a joule of electricity. With the fifth failure, even the most hard-headed nuclear engineers admit failure and the whole programme is abandoned.

On this scenario it has taken thirty-one years to learn that breeders do not work, and that the original decision to embark on the programme was mistaken. By this time, forecasts about oil costs and availability have proved correct so that there is no hope of going back to the oil which the breeder programme attempted to displace. The five breeder stations would have an output of about 45,000 GWh yr^{-1}, equivalent to about 20 million tons of coal, or 20 per cent of present electricity output. An energy shortfall of such dimensions, with no hope of alternative supplies, would be very serious.

How might this disaster be avoided? The only answer is in the provision of coal, an extra 20 million tons a year, and the means to convert at least a sizeable fraction of this to electricity, as few consumers could, in the time available, switch from electricity to coal.

The problem is that increasing coal output and building coal burning power stations takes time, so that it cannot be left until it is known to be needed because of the failure of the breeder programme. A large fossil station now takes seven years to build, so to be available when the first breeder fails, such a station must begin construction in year 8. Similarly, a second station needs to be started in year 18 in case the second breeder station fails and so on. This capacity may be made available in another way, e.g. by not closing down surplus coal burning stations, but this still costs money and must be planned for before the breeder reactors are known to fail.

There is also the time taken to increase coal production. The extra 20 million tons to be available in year 31 would be of the order of 20 per cent of present deep-mined output. Judging from the Coal Board's present plans this would appear to take 10 years if coal is mined in the usual way.[7] This could be speeded up, e.g. by beginning a number of small mines simultaneously rather than gradually developing a few large mines, but the expense of such acceleration would be large. The situation mirrors that observed in the case study on lead in petrol in Chapter 3 where new designs of car engine are to replace existing ones. At the present rate of output of engines it takes ten years to effect a complete replacement. This can be speeded up, but only at enormous cost in terms of very short lived engine plants and scrapped engines of the existing type.

To avoid the heavy cost of a crash programme for the extra coal, output must be gradually increased from about year 20 so that an extra 20 million tons a year is available after year 31 *just in case the breeder programme is a failure*. The additional coal has no value except as a hedge against the collapse of the nuclear power programme, and if the reactors work as expected then output can be gradually reduced by 20 million tons a year. If the increase and susbsequent decline in production is a linear function of time, then the coal mined as insurance would amount to 200 million tons, about two years present output. The cost of this coal and of the extra power stations to burn it must be reckoned as part of the cost of the breeder programme.

The scenario I have described is, of course, an extreme one. The failure of the various reactors happen in a particularly spiteful way, but they *could* happen like this, and the costs of such a failure would be so great that insurance must be purchased by ensuring that coal is available. Once coal production was increased in year 20, it would be more likely that the failure of the breeder design would be revealed by year 25, after which the extra coal would be needed and valuable, so that only five years of increasing production would have been used as insurance. Nevertheless, the point remains, and is best made in a qualitative way.

Breeder reactors have very long lead times, and so have a very long monitor's response time, which could be several times the lead time for an individual reactor. Error costs are failure to provide sufficient energy and are very large. The only control available, increasing coal production, has a very long lead time as well; formally, the corrective response time is very long. The system of breeder reactor plus control from coal output is, therefore, enormously inflexible. So much so that control must begin before it is known to be needed. This, of course, means that control costs are very large.

In the real world, there are many alleviating factors, such as reserves of open cast coal or gas which can be mobilized rapidly if energy output drops unexpectedly somewhere else, and partial failure, such as running at less than planned capacity, is much more likely than the total failure painted in the scenario. Nevertheless, in the real world a breeder programme must be accompanied by insurance against its failure, and given the slow response of deep-mined coal production, this is bound to have to be bought very early and is bound to be expensive; a cost which must be put down to the breeder programme.

In a way the argument merely highlights the often forgotten point that the cost of the breeder programme is the cost of the entire energy system with the programme minus the cost of the best entire energy system which has no programme. In the first case, uncertainties about the working of the new technology mean that insurance must be bought, with the cost of increasing coal production; in the latter case this is not necessary. Thus insurance costs fall to the breeder programme where they may eventually make a significant part of the programme's total cost.

References

1 Central Electricity Generating Board (1979) states a figure for the cost of AGR electricity, but has been heavily criticized for not giving the calculation on which this is based.
2 For details see Simpson (1977).
3 Department of Energy (1975).
4 Hunt (1977).
5 Hunt (1977).
6 Energy Policy (1978), p. 86.
7 National Coal Board (1977).

Bibliography

Central Electricity Generating Board (1979), *Annual Report*.

Department of Energy (1975), *Submission to the Royal Commission on Environmental Pollution, Study of Radiological Safety*.

Energy Policy (1978), *Energy Policy – A Consultative Document*, Cmnd. 7101, HMSO.

S. Hunt (1977), 'Fuel Recycling', in J. Forrest (ed.), *The Breeder Reactor*, Scottish Academic Press, 53–63.

National Coal Board (1977), *Coal for the Future*.

H. Simpson (1977), 'The International Scene', in J. Forrest (ed.), *The Breeder Reactor*, Scottish Academic Press, 37–48.

7. SCALE

Where a choice between technologies has to be made under ignorance the theory of Chapter 2 tells us to favour technologies which are easily controlled. Intuitively size and ease of control are opposed; small is controllable. One aspect of this has been discussed in Chapter 3 where it was seen that the size of some technologies, such as motor-car transport, is a serious obstacle to changing them in any fundamental way. The present Chapter considers another aspect of scale, the unit size of technologies. Economies of scale are often thought to favour very large technological units, but these can be extremely difficult to control and manage, especially where there is extensive ignorance about future input prices, processing costs, demand for output and the interactions which are possible within a complex system. Unfortunately, it often happens that the economies of scale are more easily recognized than the diseconomies which arise from difficulties in control, so that many technologies have been suspected of being on too large a scale.[1] An example which figures in the case study to follow is the British electricity generating industry.

Table 7.1 – Behaviour of a System of N Production Units

No. of failures	*Output*	*Probability of occurrence*	*Earnings*
0	Q	$(1-p)^N$	$Q(r-v)-f$
1	$(1-\frac{1}{N})Q$	$Np(1-p)^{N-1}$	$(1-\frac{1}{N})Q(r-v)-f$
2	$(1-\frac{2}{N})Q$	$\frac{N(N-1)}{2}p^2(1-p)^{N-2}$	$(1-\frac{2}{N})Q(r-v)-f$
3	$(1-\frac{3}{N})Q$	$\frac{N(N-1)(N-2)}{2.3}p^3(1-p)^{N-3}$	$(1-\frac{3}{N})Q(r-v)-f$
.	.	.	.
.	.	.	.
.	.	.	.
N	O	p^N	$-f$

Before looking at the real world, however, the intuition that ease of control diminishes with unit size needs to be examined. The simplest way of doing this is to construct a hypothetical problem where a profit maximizing firm has decided on what output to achieve, the remaining decision being whether this is best done by a number of small production units or a few large units.[2] Assume that the economics of operating units of both sizes is identical with fixed costs f; variable cost per unit output v; and revenue per unit output r and an output of Q. Profit is obtained from a unit only when it is working, so let the probability of failure of a unit be independent of its size and denote this by p. Table 7.1 lists all the possible failures in a system of N units, each with an output of Q/N, and gives the probability of each failure and the earnings which would accrue to the firm.

Table 7.2 – Behaviour of a System of Large and Small Units

No. of units in system	*No. of failures*	*Supply*	*Probability of occurrence*	*Earnings*
	0	Q	$(1-p)^3$	$Q(r-v)-f$
	1	$\frac{2}{3}Q$	$3p(1-p)^2$	$\frac{2}{3}Q(r-v)-f$
	2		$3p^2(1-p)$	$\frac{1}{3}Q(r-v)-f$
	3	0	p^3	$-f$
1	0	Q	$(1-p)$	$Q(r-v)-f$
	1	0	p	$-f$

The expected earnings by the firm can be calculated from Table 7.1.
Expected earnings

$$E = Q(r-v)(1-p)-f$$

This shows the expected earnings to be independent of N, the number of units employed. It would seem, therefore, that small units do not, after all, confer any advantage. This is not, however, the end of the matter for even with the strong assumptions above a difference can be detected between big and small units. It lies in the variance of the earnings, which is less for small units. This is easily seen from Table 7.2 where the performance of three small units and one large unit are compared. The expected earnings are as before, but the difference in the variance of expected earnings is clear.

There is little value in trying to quantify the point, as it is best made qualitatively. Under the very restrictive assumptions of the above example the expected earnings are independent of unit size, so that in the very long term there is nothing to choose between meeting output with large or small units. In the shorter term, however, the quality of the firm's cash flow is important and this favours small units. What this means is that the error cost, the cost of the failure of units, is lower for small units. This means, according to the measures developed in Chapter 2, that a system of small units is more easily controlled than one of large units. It is, to put it in other ways, more flexible, less sensitive to error and mistakes can be corrected more easily. The effect detected here may be called the *earnings effect*.

It will be worthwhile to extend the hypothetical case a little more. The profit from a system is reduced if the capacity of the system does not match demand. Undercapacity results in lost sales and overcapacity means using capital in a non-productive way. It may be imagined, therefore, that a firm aims to match its output to demand as closely as possible, and that future demand can only be forecast in a very rough and ready way, making decisions about the capacity of the system ones under ignorance. Demand is met by adding or removing a number of production units. These may be large with a long lead time for adding new units or closing old ones, or small with a short lead time. In either case action must be based on a forecast of the demand one lead time into the future. As lead time lengthens, the accuracy of this forecast will decline, with a consequent increase in the mismatch between capacity and demand. Once again, the conclusion is that small units of low lead time are more easily controlled than large units with a long lead time, and we may say that this is due to the *forecasting effect*.

A third source of flexibility for a system of small units lies in what may be called the *step effect*. It arises where demand changes continuously but capacity varies stepwise, as in the example of the last paragraph. If C_S and C_L are the capacities of a small and a large production unit, then there are more numbers equal to $n\ C_S$ (n is an integer) than numbers equal to $n\ C_L$. This means that there are more options about the system's capacity when small units are used, which tends to reduce error cost, measured by the mismatch between capacity and demand. This may be clearer from Figure 7.1.

There is yet a fourth effect, the *control effect*. To establish its independence of the forecasting and step effects, imagine that the firm has a choice between units of the same capacity but different lead times, and that the

Figure 7.1
Effect of Size of Production Unit on Ability to Follow Changing Demand
(shaded area is mismatch between demand and capacity)

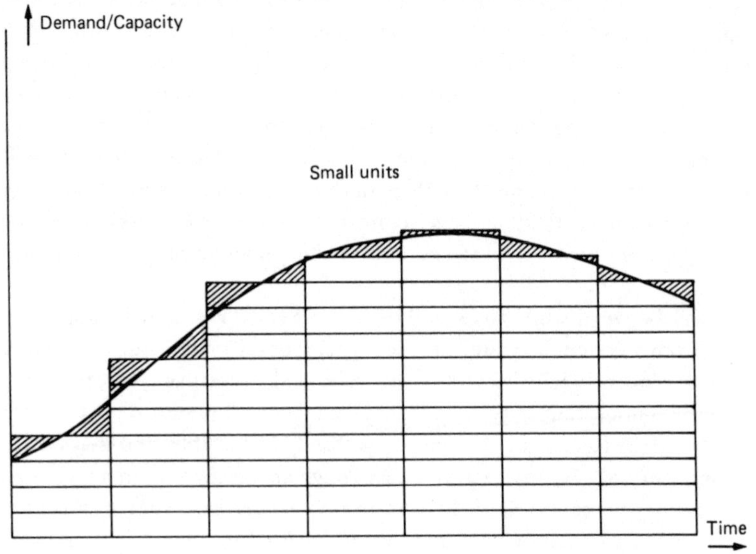

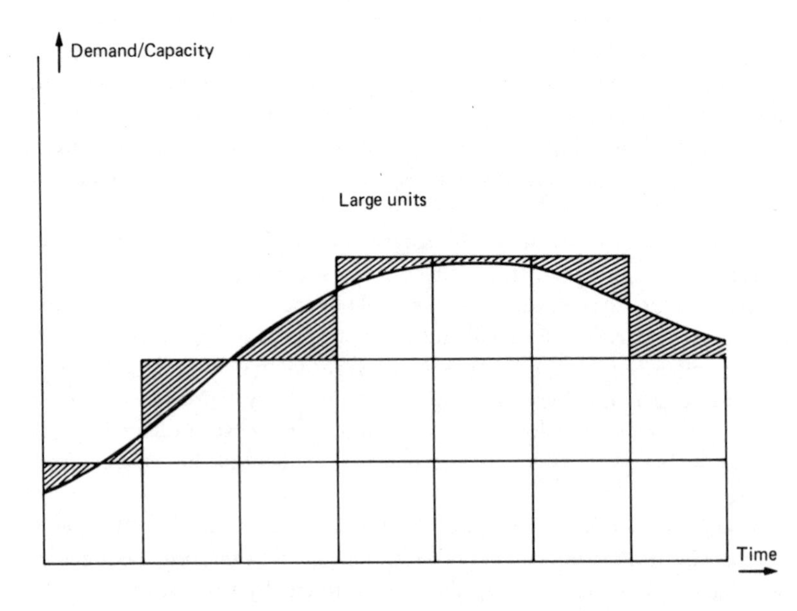

quality of forecasting over the short lead time is the same as over the long lead time. The second assumption amounts to the probability distribution of the forecast error being independent of the period of the forecast. This, of course, is a highly unlikely fiction, but it serves our present purposes. The control effect arises because the system with a short lead time can respond more quickly to information that a forecast has been wrong. The error can be eliminated more quickly than in the case of the system whose units have a long lead time, so giving a smaller mismatch between demand and capacity, or a smaller error cost. If the demand proves to be higher than forecasted, so that extra capacity is required, this can be added more quickly using units of low lead time. Similarly if demand is less than forecasted. The mismatch between demand and capacity is, therefore, lowered by reducing the lead time of the production units.

A fifth effect, the *learning effect*, also enhances the flexibility or ease of control of small units. If a number of small units are constructed instead of a few large ones learning about the units' behaviour and reliability is likely to be more rapid, even if the lead times for both units are the same. The learning is even more rapid if the small units have a lower lead time. A system of small units is, therefore, likely to operate more efficiently than a system of large units, resulting in lower error costs, and greater flexibility. The importance of learning will differ greatly from one system to another, so that little can be said at a general level.

It may be concluded, therefore, using the measures stated in Chapter 2, that a production system of small units with short lead time is more controllable and more flexible than a system of large units of long lead time. Decisions about the former are more corrigible, and its performance is less sensitive to forecasting errors. Control costs, the costs of buying and operating new units and scrapping old ones, will, however, generally favour large units because of economies of scale. This is to be expected, as flexibility usually has to be paid for in some way. The essence of the sort of decision making under ignorance discussed above is the trading between flexibility and control cost.[3]

Case Study – Electricity Generation

The discussion of the artificially simple problem above may now be given a little depth by considering the electricity generating industry and the unit scale of its operation.[4] Until 1954 almost all new generating plant was of 30 or 60 MW units, but after this there was a very rapid increase in unit size, stemming from the belief that large economies of scale existed. Table 7.3 shows the pace of this development.

Table 7.3 – The Development of Generating Plant

Year	*Rating (MW)*
1950	100
1951	100
1952	120
1953	200
1956	275
1957	550
1958	300
1959	350
1960	375
1960	500
1966	660

Between 1965/66 and 1974/75 the Central Electricity Generating Board introduced six 350 MW units, two 375 MW units, two 660 MW units and forty-five 500 MW units. The case for such scale was fourfold. Large plant would enable the expected very rapid increase in demand to be met; capital cost per unit output is lower for large plants; large plant has a higher operating efficiency because leakages and heat and power losses fall in relation to the total steam flow; and unit costs of maintenance and labour are lower for large plants.

Against this, there are a number of diseconomies of scale, some of which illustrate the various effects discussed above and in previous Chapters.

(a) Large power stations take longer to build than small ones, so that errors in the forecast of the capacity which will be required are greater. Fossil-fuelled stations of the size now regarded as optimum by the CEGB take 7-10 years to build and nuclear stations 10-14 years. The forecast effect here is, therefore, likely to be very serious. In such circumstances, according to Patterson:

> Ordering the station then becomes not an act of foresight but an act of faith, founded on policy. The suppliers must then endeavour to ensure that electricity demand 10 years hence will have increased to the level warranting addition of the new station. The internal objectives of the system planners thus take precedence over the social and economic role of the system. Planning, as conventionally understood in mixed economies, disappears; the technology takes over and reproduces itself according to its own introverted criteria.[5]

Table 7.4 – Forecasting Error in Five Year Forward Estimate of Maximum Peak Demand 1960-1970

Year	*Forecasting Error %*
1960	-4.0
1961	+3.4
1962	+11.5
1963	+24.0
1964	+28.5
1965	+28.8
1966	+27.3
1967	+25.0
1968	+24.0
1969	+22.6
1970	+24.0

Table 7.4 gives the record of forecasting the five year ahead maximum peak demand, expressed as a percentage of the estimate.[6] The year 1960 marks a transition from small, consistently negative errors since 1947, to much larger positive errors. One reason was a change in forecasting technique around this time.

(b) The long lead times of large plant mean that forecasting errors are compounded because they cannot be corrected quickly. This is, of course, the control effect. This was a problem before 1965, when the tendency was to underestimate the capacity required, but in recent years there has been considerable overcapacity. The time needed to close unwanted plant is, of course, much less than that needed to add capacity to the system.

Because forecasts are very unreliable and, if underestimates of demand, very slow to correct, the planning margin of the generating system is very large. This is the capacity additional to that forecasted as required to allow for errors in the forecast and unexpected failure of equipment. As plant has become larger, the margin has increased from 14 per cent of forecast capacity, to 17 per cent, reaching 20 per cent in 1968 and the present 28 per cent in 1977, and it is likely to achieve the remarkable 35 per cent in the near future. Because the margin represents capacity that is unlikely to be used, it imposes a considerable cost on electricity generation.

(c) It may be illuminating to inquire why the system has developed a very large excess capacity in recent years, as shown in Figure 7.2.[7] I do not mean that we should delve into the detailed reasoning of the forecasters in the CEGB Planning Department, but that we merely ask why the system has tended to be planned to conservatively over this period. The reason is to be found in Chapter 3 where the behaviour of highly valuable low variety systems was discussed. The electricity system, which includes the generating plant, transmission and all electricity users, is a low variety system. Capital equipment designed to use electricity cannot generally use any other form of energy, so that in the short term electricity is essential for the operation of a vast range of equipment. Virtually all domestic and about 85 per cent of industrial electricity is supplied from the grid. The cost of a failure in supply is, therefore, very high, for there is no short-term substitute. The planners of the supply system, therefore, have to hedge so that they will be able to provide electricity even if demand is much higher than their forecasts indicate. Patterson strongly objects to the inflexibility that the need to avoid this large error cost imposes on the way the system is planned:

Figure 7.2
Declared Net Capability of CEGB Plant (A) and Maximum Demand (B)

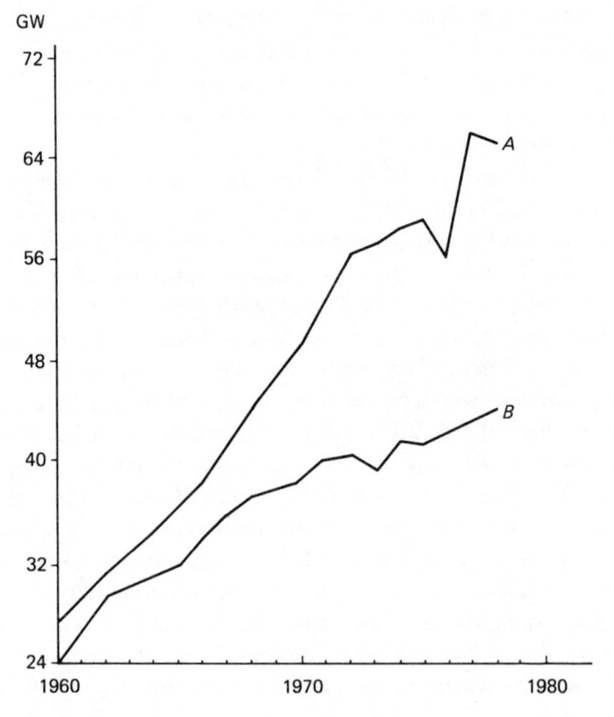

> The nature of grid electricity as an essential commodity removes the ultimate sanction of bankruptcy in the event of unfulfilled plans. Public participation in planning is at least an inconvenience; public opposition to particular plans – say for the siting of a new power station – may become such an inconvenience that it may have to be administratively overruled.[8]

(d) Large units have a low availability when they are first commissioned. Part of the reason is that learning about the operation of a big plant is more difficult than a small plant. This is particularly serious because the early years are the most significant in the calculation of the unit's cost.
(e) The step effect has no significance for a system as large as the CEGB's, but its role in planning smaller systems has been noted by Schroeder and Wilson (1958). Ignoring the other effects, they considered the effect of commissioning a single unit, whose size is a significant fraction of the system's total capacity, on the load factor of other plant. The new plant will take the system's capacity above the demand, so that older units will have to operate at lower load factors. This is a cost, since it results in idle capital equipment. They argued that in a small system the best size for generating units was 150–250 MW, 300–400 MW units being substantially more expensive.

There are other factors which lead to diseconomies of scale, but the above are the ones relevant to a general discussion of the effect of unit size on the ease of control of a production system. One factor which is very common, though not universal, and so worthy of mention is that large generating units have an inherently lower availability than small ones. For example, a 500 MW boiler will have ten times the tubing of a 50 MW boiler, and so with the same quality control, can be expected to have ten times the downtime.

With these diseconomies of scale in mind Abdulkarim and Lucas (1977) set out to determine the optimum size of fossil-fuelled generating plant for the CEGB's system. The existing system was compared with four hypothetical systems of unit size 100, 200, 300 and 400 MW, with lead times of three, four, five and six years. Capital costs for each type of unit were calculated, and estimates made for fuel and non-fuel operating costs. For each year between 1965 and 1976 each hypothetical system was made to follow that year's load-duration curve so the operating time of each unit could be found. From this the operating cost of the whole system was derived. The present value of the capital and operating costs discounted at 10 per cent to 1965 was then calculated, with the results given in Figure 7.3. There is a sharp minimum between 200 and 300 MW.

The authors stress the strength of the assumptions they have had to use, and propose a qualitative view of their results. What the study shows is that the trade-off between economies and diseconomies of scale for the CEGB's electricity generation system gives an optimum unit size of between 200 and 300 MW.This is far smaller than present thinking in the industry maintains, where it seems to be held that the bigger the better, at least for generating sets up to 660 and even 1,300 MW.

Figure 7.3. *The Present Worth in 1965 of the Operating and Capital Costs of Electricity Generating Systems of Different Unit Size.*

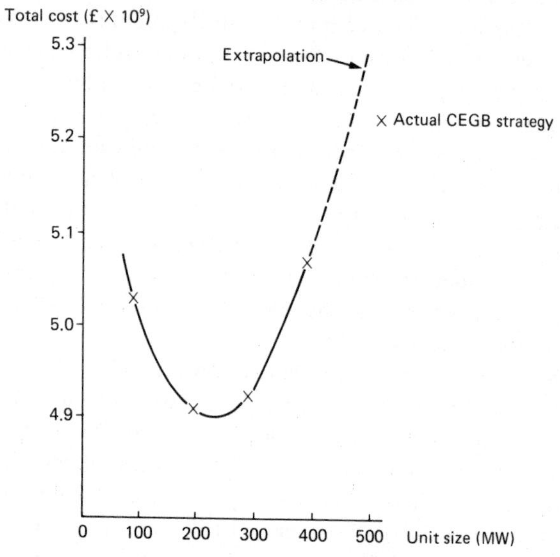

Before leaving the study it must be noted that it is essentially *post hoc*, using the actual load duration curve for each year and historical costs. Decision makers at the time could not have used the method to determine an optimum plant size. A remarkable thing about the study is the time it was undertaken; about fifteen years after the extensive building of units which were too large. The assessment of economies and diseconomies of scale turns out, surprisingly, to be extremely difficult, even when all the operating data from plants of different size is available.[9] If mistaken beliefs about these are adopted and become institutionalized, as in the CEGB, it can be a very long time before an analysis emerges which is sufficiently sharp to correct the error. In the meantime, of course, the costs from building plants of the wrong size is gigantic.

Given the scepticism expressed about forecasting in Chapter 1 and elsewhere, it may be conducive to a more balanced understanding to conclude the case study with a discussion of the role of forecasting in planning the electricity system. The simple-minded view which seems to be taken by the CEGB is that the cheapest electricity is provided by a unit of such and such capacity; units of such and such capacity need so many years to build; hence forecasts of demand must be made so many years ahead. If forecasting errors produce a system of the wrong capacity, the fault lies with the methods used which must be improved. The problem should be seen in quite another way. The CEGB is in control of a system, the demand for whose output can only be forecast in a very rough way. Since forecasts will generally be in error, a system where forecast errors can be easily corrected and whose performance is, therefore, not highly sensitive to such errors, should be developed. This may well entail exploiting the shorter lead times of units smaller than those able to produce the cheapest electricity. A system of such units will perform better than a system of larger units; the mismatch between capacity and demand will be smaller. On the other hand, generating costs will be greater due to loss of economies of scale, assuming their existence. The controller of the system must, therefore, come to some compromise between ease of control and insensitivity to forecast error and generating costs. The present automatic increase of forecast capacity in the operation of the planning margin to allow for forecast error is far too crude a device. It may be effective, but this needs to be demonstrated against other ways of coping with error which employ units of smaller size.

The above discussion reveals the tension which exists in many decision-making problems under ignorance, between the ability to control the system in question and the need to predict its future behaviour. At one extreme, if it is possible to predict with confidence satisfactory future performance of the system, then there is no need to ensure flexibility. Flexibility, after all, is only required when things go wrong, so if it is known that the system's future behaviour will be satisfactory, there is no call for flexibility. At the other extreme, if a system has the highest possible flexibility, so that it can respond to changes with zero error cost, and if its control costs are also zero, then there is no need to predict how it will behave. It can always be adjusted to an acceptable state at zero cost. All real systems, of course, are between these extremes. What this means is that flexibility must be traded off against the need to forecast the system's future behaviour. This is the familiar trade-off between controlled error cost and control cost in a different guise. The performance of a flexible system will be less sensitive to errors in forecasting than that of an inflexible system, since it bears lower controlled error costs, but it will generally bear greater control costs. The decision maker must, therefore, compare an inflexible system with a high sensitivity to forecasting error, but low control cost, with a flexible system of low sensitivity to forecasting error, but high control costs. Where the balance will be will depend upon the difficulty in forecasting the system's behaviour. Where this is a serious problem, an appropriate degree of flexibility must be purchased.

References

1 For another cause of excessive size see Chapter 5.
2 The following discussion follows Ball (1975).
3 This is not to say that an optimum trade-off can be identified – see the discussion about buying a car in Chapter 2.
4 The following discussion follows Abdulkarim and Lucas (1977).
5 Patterson (1977), p. 49.
6 From Abdulkarim and Lucas (1977).
7 From Electricity Council (1971-1977); Electricity Council (1978/79).
8 Patterson (1977), p. 91.
9 Fisher (1978) and Lee (1978).

Bibliography

A. Abdulkarim and N. Lucas (1977), 'Economies of Scale in Electricity Generation in the UK', *Energy Research, 1*, 223–31.

D. Ball (1975), 'Flixborough – Too Many Eggs in One Basket?', *Process Engineering*, August, 55–9.

Electricity Council (1971–1977), *Handbook of Electricity Supply Statistics*.

Electricity Council (1978/9), *Statement of Accounts and Statistics*.

J. Fisher (1978), 'Economies of Scale in Electric Power Generation', International Institute for Applied Systems Analysis Seminar, 10 October, IASSA Vienna.

T. Lee (1978), 'Optimization of Size in Power Generation', International Institute for Applied Systems Analysis Seminar, 5 December, IASSA Vienna.

W. Patterson (1977), *The Fissile Society*, Earth Resources.

T. Schroeder and G. Wilson (1958), 'Economic Selection of Generating Capacity Additions', *Transactions of the American Institute of Electrical Engineers*, 77, 1134–45.

8. DOGMATISM

The last two Chapters may have given the impression that all obstacles to the control of a technology are determined by the technology itself; by its scale, unit scale, lead time and so on. This would be a mistake, for human beings and their social organizations often operate to actively hinder the control of technologies in which they have an interest. If an organization wants to develop a technology for its own ends, it is notoriously difficult to convince its members that this will be a bad thing, and groups dedicated to thwarting some technological adventure are often just as hard to convince about the value of the technology which they oppose. The essence of decision making under ignorance is the exposure of a decision, once made, to falsification from facts, and the maintenance of the ability to correct the decision should it be falsified. But it was observed in Chapter 2 that there are a number of face-saving devices which proponents of the option which appears to be falsified can employ to prevent its falsification. Groups with favoured technological plans can exploit a whole armoury of such protective dodges to deflect criticism of the options they support. This Chapter considers these *ad hoc* moves, the making of which amounts to the dogmatic refusal to accept criticism of cherished options. A far from complete list of these moves follows:

(a) Ignoring Falsification

Decisions on the building of major roads in Britain are made by the Department of the Environment and based on forecasts of the traffic which will use the road. Until the severe criticisms of Leitch (1978), the Department did not bother to check its forecasts against actual road use by simply counting traffic. This would have revealed errors in forecasting which the Department might have found embarrassing. Although the roads involved could hardly be ripped up, systematic errors in the forecasting methods of the Department could have been corrected had they been detected.

(b) Cooking the Books

Lord Hinton tells the following story about his involvement in the decision to build the United Kingdom's first experimental breeder reactor at Dounreay:

> Because of the unknowns we had, from the outset, planned to build the reactor on a remote site. Kendal, Owen and I had examined the Galloway and South Ayrshire coasts without finding anything that was suitable. We decided that we had to go further afield and surveyed the coast of Sutherland and Caithness and there we found what we knew to be the best site at Dounreay. But why, if we were giving the reactor containment, were we putting it on a remote site? This could only be logical if we assumed that the sphere was not absolutely free from leaks. So we assumed, generously, that there would be 1% leakage from the sphere, and dividing the country around the sites into sectors, we counted the number of houses in each sector and calculated the number of inhabitants. To our dismay this showed that the site did not comply with the safety distances specified by the health physicists. That was easily put right; with the assumption of a 99% containment the site was unsatisfactory so we assumed, more realistically, a 99.9% containment and by doing this we established the fact that the site was perfect.[1]

(c) Inability to Monitor

Nothing could be easier than counting traffic to monitor a decision to build a road, but one way of avoiding the discovery of mistakes is to make decisions which cannot be monitored so easily, or cannot be monitored at all. All western governments spend very large sums in all sorts of ways in an attempt to stimulate technological innovation by their industries, but very little has been learned about whether these efforts have any effect on the rate of innovation. In a review of these government attempts Braun (1980) concludes: 'Government has a large range of measures in its armoury by which to influence different aspects and stages of innovation. The efficiency of the various measures is, however, largely unknown.'[2]

The problem is not so much refusal to examine the effectiveness of the measures taken, as in the Department of the Environment case, but that their examination is just extremely difficult. Thus, the decision makers involved can continue to operate the programme they favour without living in fear that it will one day be shown to be wrong.

(d) Imprecision

If the decision maker's objective is stated very imprecisely, then it is very difficult, often impossible, to ever point to events which reveal his decision to have been mistaken. Many political decisions are of this sort. What, for example, was the Government's objective in joining the EEC – what did they do it *for*? At the time the objectives of the Government were very poorly articulated, so it is hard to point to events which show them to have been wrong. Does the outflow of funds from Britain through the EEC agriculture support system show them to be wrong? It might do, if the objective was to improve the British balance of payments – but was it? Does the poor performance of British industry show that joining was wrong? It might do, if the objective was to improve British productivity. But was this the objective?

An interesting technological example here is the third London airport, in particular the Report of the Royal Commission under Justice Roskill in 1971. The Commission's task was to advise on a site for a third London airport and the timing of its construction. Very little was said in the Report about the objectives which the new airport was supposed to meet. Obviously it was supposed to accommodate future growth in air traffic to and from the Metropolis, but exactly why this was a worthwhile aim is very unclear from the Report. The fear was that:

> congestion in the London area almost certainly would lead to a shift of traffic to competing continental airports such as Paris, Amsterdam and elsewhere and hence to competing foreign airlines. These cities were anxious to acquire for new airports already under construction the traffic which airports in the London area could not absorb but which they could. Thus this country was already in danger of losing its position as a great centre of world air transport contrary to what had been recommended in the White Paper of 1953. Heathrow was handling in 1968 more international traffic than any other airport in the world and more than that handled by the airports of Paris and Amsterdam combined. The phrase 'London is the Clapham Junction of the air' was no mere chauvinistic cry. It represented the deeply and sincerely held views of the economic and political importance of maintaining the position of this country as one of the two foremost aviation nations of the Western world and as the leading aviation nation in Western Europe. The hostile jibe during the Second World War that this country was no more than an aircraft carrier should in the last thirty years of the present century be a source not only of pride but of economic and political strength.[3]

Later on, the need for the new airport to support tourism is stated:

> If the third London airport site is less attractive than other available airports, travellers will not use it. Demand will build up elsewhere either in this country or even overseas. The American promoter of package tours will cease to include this country on the itinerary of his European tours if the costs and difficulties of getting the tourists to and from the airports become a deterrent; when Paris, Rome and Athens are in the brochure but not London he will be a determined Anglophil who forgoes the journey because London is not included in the intinerary.[4]

Is it really necessary to state the objectives of the airport any more fully and precisely than this; is not the growth in air traffic enough to justify its construction? Unfortunately, a good decision does require a better statement of objectives than the ones given here, because the building of a third London airport may not be the most effective way of meeting these objectives. If the airport is required to maintain the British aircraft industry it is pertinent to ask in what other ways could this be achieved and with what effectiveness and what cost. If the new airport is supposed to maintain London's tourist industry, then what other ways of doing this are available, and are any better than building the airport? Given the huge cost of the proposed airport, there might well be cheaper and no less effective ways of meeting its objectives. This, of course, can only be settled if these objectives are well defined. Leaving them in the imprecise form thought satisfactory by the Roskill Commission means that it is very difficult to criticize the proposed airport by showing that its objectives are better met in some other way. In asking where the third London airport should be sited, the principal question of whether this is really the best way to do whatever it was attempting to do has been glossed over.

(e) Waiting for Vindication

Here, whatever bad effects a decision is found to have it is held that this will be outweighed by future benefits. This is used by politicians in an almost reflex way, but is particularly barbed when used by revolutionary leaders who assure their followers that the pain of today's purges and collectivizations will be outweighed by the joys of the just society once it is achieved.

The decision by the Central Electricity Generating Board to build their Advanced Gas-cooled Reactors (AGRs) must rank as one of the most disastrous technological decisions in post-war Britain. As mentioned in Chapter 6, these took far longer to construct than expected, Dungeness B being still unfinished after fifteen years; the performance of those which have been finished has been nearly the worst of all reactors in the world and the electricity demand which they were supposed to meet, by great fortune, never materialized, and yet it is still maintained by their constructors that the decision was a good one because the AGRs will soon be providing electricity at a much lower cost than that from coal.

(f) Changing Objectives

The objectives of a decision maker can change so as to protect a decision he has made in the past. The original objective in the decision to develop Concorde was the saving of the British aircraft industry. When it became clear that this was not going to be achieved by building Concorde, the project was not abandoned, but the objective was changed to one of making profit for the British Aircraft Corporation. When it was realized that Concorde would make little, if any, profit, its objective was changed once again, to that of obtaining French co-operation in Britain's attempt to enter the EEC. When this was not forthcoming, the project was still not cancelled and its proponents argued that the plane would make a profit after all, albeit over a longer period than originally expected. Thus, we are blessed with supersonic passenger transport. This is an important feature of the case study which follows.

(g) Catch 22

Here things that would show the decision maker to have been wrong are simply not allowed to happen. In the famous case of Catch 22, the US Army decided that mentally unstable soldiers could apply for release from their duties. This would have been wrong, according to the Army's objectives, if it lead to a flood of soldiers leaving the Army. It was, therefore, decided that any soldier with the wit to apply for release was sane enough to continue in the service. In this way no-one was released and the military decision makers were spared the embarrassment of having a mistake revealed.

The suggestion that nuclear power should be developed only when it is shown to be safe is in the same category, because it amounts to a total prohibition of nuclear technology, proof of safety requiring operating experience of nuclear plant. In a more serious vein, examples which come very close to Catch 22 were considered in Chapters 4 and 5. MIRV was needed by the United States as a hedge against various developments of the Soviet Union's deterrent forces, but the acquisition of MIRV ensured that the feared development actually happened. The American decision to deploy MIRV was virtually certain to be vindicated in this Alice in Wonderland way, so that no military planner needed to lose sleep about the fears which had prompted the decision proving to be groundless. In the same way, fears about the adequacy of future energy supplies leads to hedging in the form of very generous provision in the future, this generosity mitigating against the efficient use of energy and realizing the feared high demand.

In making decisions under ignorance, decisions should be open to falsification from facts, and so all of these dogmatic devices for evading criticism of a favoured option must be carefully guarded against. The problems of such vigilance are well illustrated by the case study which follows.

Case Study – The Manhattan Project

The story of the invention of the atomic bomb and its use against Japan is too well known to require detailed recounting here. Rather than the whole elaborate story, our concern is with a tiny part of the whole, which was, however, crucially important to the development of the post-war world. Despite this importance, the issue has, as far as I know, received no critical discussion until now. The issue I wish to focus on is the change in the objectives of the Manhattan Project from a defensive reaction to the possibility that the bomb would be made in Germany, to the provision of an offensive weapon against Japan; there being no suggestion that this country might obtain a bomb.

In the spring of 1940, Otto Frisch and Rudolf Peierls, two refugee scientists in Birmingham, wrote their now famous memorandum, arguing that a superbomb with an explosive power equivalent to some thousands of tons of TNT could be made from suitably processed uranium. The authors of the memorandum were modest enough to countenance the possibility that scientists working in Germany had reached this same conclusion, and that work on the bomb's construction might, therefore, be underway within that country. The fourth conclusion of the memorandum is worth quoting:

> If one works on the assumption that Germany is, or will be, in the possession of this weapon, it must be realized that no shelters are available that would be effective and could be used on a large scale. The most effective reply would be a counter-threat with a similar bomb. Therefore it seems to us important to start production as soon and as rapidly as possible, even if it is not intended to use the bomb as a means of attack. Since the separation of the necessary amount of uranium is, in the most favourable circumstances, a matter of several months, it would obviously be too late to start production when such a bomb is known to be in the hands of Germany, and the matter seems, therefore, very urgent.[5]

The Frisch-Peierls memorandum found its way to the Committee on the Scientific Survey of Air Defence and Air Warfare, where it aroused great interest. The Maud Committee was soon set up to produce the weapon as rapidly as possible, a task which was undertaken with so much energy that the Committee is often seen as the most successful in British history. By the summer of 1941 the Committee reported that a U-235 bomb was certainly feasible, as was probably the case for a plutonium weapon. The question then began to shift from if the bomb was possible to where and how it should be made.

In the USA, President Roosevelt was warned in October 1939 by another pair of refugee scientists, Szilard and Einstein, that Germany might be working on an atomic bomb. As a result some work was done on the possibility of the suggested bomb, but in a much more leisurely way than in Britain. It must be remembered, of course, that the United States was still neutral. In the summer of 1941, however, the report of the British Maud Committee was sent to the United States, which poured life into the previously flagging American project. Within six months of Pearl Harbour, in December 1941, the American work had already gone further than the Maud Committee, and by mid-1942, when the title Manhattan Project was employed by the Americans, the gap between the two projects was becoming an embarrassment to the British. After protracted efforts, Churchill and Roosevelt signed the Quebec agreement in August 1943, which allowed British participation in the American project. The British project virtually ceased with migration of the bulk of its scientists to America.

Through all this time, the stated objective of Allied efforts to make a bomb was to counter any similar developments in Germany. There was little idea of the use Germany might make of such a weapon, and how exactly the possession of a similar device by the Allies would affect the military balance, but the motivation was the common enough one in time of war; one seeks to have all the weapons available to the enemy, even if their final use cannot be foreseen with any clarity. This, coupled with the great technical uncertainties, is enough to provide an excellent example of a decision under ignorance.

We have seen the need to monitor such decisions, and the British were apparently aware of this. From the beginning of 1942 British Intelligence took a great interest in all German atomic projects. They studied the whereabouts of nuclear physicists, the buying of uranium and heavy water across the world, and the construction of novel industrial plants which might be used towards the construction of a bomb. As time went on, it became increasingly clear that work in Germany was on a tiny scale compared to Allied efforts and mainly centred on slow neutron piles for energy production. Nevertheless, precautions were taken such as the spectacular destruction of German heavy water stocks and the plant for their production in Norway.

In early 1944, British intelligence experts were confident that the Germans had no large atomic projects and would not be able to manufacture a weapon in the foreseeable future. Their conclusions were reported to the American intelligence community, who were a little more cautious in their interpretation. They felt that as long as the possibility of a German bomb was above zero, the British conclusion could not be upheld. Confirmation of the British conclusion was, however, soon forthcoming, as the invasion of Germany showed the picture the British had built of German nuclear efforts to be accurate in almost all respects.

During this period the work of the Manhattan Project had continued at a frantic pace, which was in no way slowed by the intelligence findings discussed above. In August 1944 General Groves, director of the Manhattan Project , reported to the Army Chief of Staff, General Marshall, that several plutonium weapons would probably be available between March and June 1945. If unforeseen difficulties with this design were found, then a single uranium bomb would be available in August. It was clear to both, looking at the military situation in Europe, that the bomb would be ready too late to be effective there. Not only was the bomb not needed to counter a German weapon, it was not even required in the war against Germany.

By this time, there had been a gradual shift in American thinking from the need for the bomb as originally envisaged, to its use as an offensive weapon against Japan. As early as May 1943 the Military Policy Committee suggested that the first bomb be used against Japanese naval forces at Truk. Later that year, General Groves singled out B-29 aircraft, designed for the Asian theatre, for modifications for atomic bomb delivery and not aircraft operating in Europe.

In his first interview with the new President, Truman, Secretary of War Stimson urged him to establish a committee to consider the employment of the bomb. The committee consisted of five politicians and three eminent scientists, Vannevar Bush, Karl Compton and James Conant. The committee, known by the bland name of the Interim Committee, was bolstered by a scientific panel drawn from Manhattan staff of J. Robert Oppenheimer, Enrico Fermi, Arthur Compton and Ernest Lawrence. In June 1945, the Interim Committee recommended that the bomb be used against Japan as soon as possible, against a mixed military–civilian target, and with no warning or previous demonstration. Although secret, the recommendations filtered down the Manhattan Project, where it roused many senior scientists to express grave misgivings, prompting the writing of the Franck Report which was signed by seven prominent scientists and sent to Stimson. The Report foresaw with remarkable accuracy the arms race which would follow the use of the atomic bomb and which, its authors declared, could only be avoided by establishing international trust which would be jeopardized by a surprise attack on Japan. It stated that:

> from the 'optimistic' point of view – looking forward to an international agreement on the prevention of nuclear warfare – the military advantages and the saving of American lives achieved by the sudden use of atomic bombs against Japan may be outweighed by the ensuing loss of confidence and by a wave of horror and repulsion sweeping over the rest of the world and perhaps even dividing public opinion at home.
>
> *From this point of view, a demonstration of the new weapon might best be made, before the eyes of representatives of all the United Nations, on the desert or a barren island.* The best possible atmosphere for the achievement of an international agreement could be achieved if America could say to the world, 'You see what sort of a weapon we had but did not use. We are ready to renounce its use in the future if other nations join us in this renunciation and agree to the establishment of an efficient international control.[6]

But even if international control is regarded as impossible, there are still good reasons for foregoing a surprise attack on Japan. If there is no international control, the first demonstration of the new bomb will trigger an arms race, which it is in the interests of the United States to delay as long as possible in order to advance her technological lead even further:

> The benefit to the nation, and the saving of American lives in the future, achieved by renouncing an early demonstration of nuclear bombs and letting the other nations come into the race only reluctantly, on the basis of guesswork and without definite knowledge that the 'thing does work,' may far outweigh the advantages to be gained by the immediate use of the first and comparatively inefficient bombs in the war against Japan.

In August, as part of the same reaction to the Interim Committee's recommendations, Leo Szilard sent Stimson a petition signed by himself and sixty-nine colleagues arguing that the first nation to use atomic weapons might have 'to bear the responsibility of opening the door to an era of devastation on an unimaginable scale'. There were countermoves to this reaction though, and a poll of scientists at the Metallurgical Laboratory in Chicago showed that 15 per cent favoured the use of the new weapon against Japan as effectively as possible; 46 per cent called for a military demonstration in Japan with an opportunity to surrender and 26 per cent favoured a demonstration in the United States.

These pleas, however, made no impact. Stimson sent the Franck Report to the four-man scientific panel of the Interim Committee. Remembering their deliberations at a later time, Oppenheimer recalled that:

> We said that we didn't think that being scientists especially qualified us as to how to answer this question of how the bomb should be used or not; opinion was divided among us as it would be among other people if they knew about it. We thought the two overriding considerations were the saving of lives in the war and the effect of our actions on the stability, on our strength and the stability of the post war world [sic]. We did say that we did not think that exploding one of these things as a firecracker over the desert was likely to be very impressive.[7]

Szilard's petition met a similar fate, being referred to the whole Interim Committee, with as little effect as the Franck Report. The question posed to them as recalled by Compton was 'can you think of any other means of ending the war quickly?'

Whilst this was going on, the bomb had been successfully tested on July 17th at Alamogordo with a yield far higher than anticipated. On August 6th a uranium bomb was dropped on Hiroshima. The yield of the weapon was around 20,000 tons of TNT and the destruction appalling. In the city 4.4 square miles were burned out completely; and between 70 and 80,000 people were killed. About the same number were injured. Three days later a plutonium bomb was dropped on Nagasaki with the same yield as the Hiroshima weapon. Of that city 1.8 square miles were destroyed with around 40,000 dead and 40,000 injured. The following day saw the effective surrender of Japan.

Lessons from the Case Study

The element in the Manhatten Project's history which is of interest to us at the moment is the change in its objective. What began as an attempt to counter the German production of an atomic bomb became the construction of an offensive weapon for use in the war against Japan. Any change in the objective of a technological project like the Manhattan one can be seen in two extreme ways. The technology's proponents may be seen as determined to preserve the project come what may, so when it is found not to serve its original objective an objective is invented which it does serve. At the opposite extreme, the change in objective may be viewed as a perfectly healthy exploitation of possibilities opened as the project proceeds. There is, of course, a spectrum between these two end points. What we may ask, therefore, is which of these extremes is nearest the truth in the case of the Manhattan Project? Thus our interest is not in the merits of the change of objective, but in the motivation behind it: was it from a dedication to find some use for the new marvel of technology, or did it result from a cool and realistic appraisal of possibilities which had not been considered until opened in the course of the project? In discussing the particular case, we shall perhaps learn one or two lessons of a general nature about the problems of controlling technology.

The following points need to be kept in mind in assessing the nature of the change in the Manhattan Project's objective.

(a) The change in objective called for no changes in the nuts and bolts of the Project's work. The identical hardware could be used to match a German atomic bomb, or to kill Japanese. For this reason the change in objective occurred gradually and insidiously. There was at no time, therefore, a pressing need to determine what the objective of the project should be, and, in fact, no real discussion ever took place. The agonizing towards the end of the Project mentioned above was about the question 'now we have a bomb should we use it against Japan, and if so how?'. But at no time was consideration given to the much more fundamental question, 'now we do not need the bomb to counter German developments, should we have a bomb at all?'. This should have been a question for debate from mid-1943 to, with increasing urgency, 1944.

(b) British Intelligence formed the very strong opinion that the German threat was unreal by 1944, but American Intelligence continued to be sceptical of the British conclusions. This they could afford to do because no decision rested on the intelligence interpretation. If the British were correct, the bomb could be used against Japan; if they were mistaken, the bomb would be needed primarily to counter German developments, although any surplus might be used against Japan. In either case, the Project's task was to make an atomic bomb. Thus the Americans could afford to doubt British thinking. If a major change of resources were required for a change from a German to a Japanese weapon, no doubt the Allies' intelligence agencies would have come to a rapid agreement.

(c) There was no demand from the military for an atomic bomb to improve the efficiency of the air attack on Japanese cities, for the existing technology was very impressive, achieving casualty rates higher than those at Hiroshima and Nagasaki. Low level night fire raids, begun in early March

1945 were particularly terrible. The first of these, against Tokyo, destroyed a quarter of the city's buildings, making one million people homeless, and leaving 80,000 dead. Thus the defeat of Japan in general, and the strategic air offensive in particular, in no way depended on an atomic bomb. At most, the bomb saved American servicemen; but this is a far cry from the situation the fear of which had led to the whole Project. If Germany had an atomic bomb, then there was a good chance that the only way of ensuring her defeat was for the Allies to have a similar weapon.

(d) Given that the atomic bomb was not essential to Japan's defeat, as it had been feared essential to Germany's, the question must be asked whether the resources invested in it after late 1943 – early 1944 might have been better spent on other ways of waging war against the Japanese; particularly when the resources included the finest group of scientists in the world. It is a little late for an answer now, but the question was never seriously debated at the time.

(e, If those in charge of the Manhattan Project were convinced that its product was not required to counter a German weapon, and were seriously considering whether to end the Project, a key issue would have been the benefits obtainable from diverting the huge effort already made for military purposes into civil nuclear technology. It may have been greatly to the good of the United States to begin the post-war period with as large a lead as possible in what was seen as potentially a vital technology, and military domination of research might have reduced this lead. Whatever the truth of the matter, the question was never discussed.

(f) The change in objectives was not morally neutral. It is one thing to develop an horrendous weapon when one is forced to do so because it may be in the hands of an enemy; it is quite another to build such a weapon for use against an enemy who does not possess it himself. Although such questions were certainly discussed by some of the Project's scientists, they received no airing in more elevated political circles.

(g) Even though the destructive power of the new weapon was greatly underestimated until the first test, it was apparent to many scientists in the Project and to politicians like Stimson, American Secretary of War, and to Lindemann in Britain, that the only way to control the new technology was by some kind of international organization, perhaps based on the free interchange of scientific ideas. They realized that the new weapon would not fit with the traditional nationalism of political bargaining and threat. The development of the first bomb may well have been very significant here. Maybe the Franck Report was right in thinking that control would be more difficult once the bomb had been made. When it is known that the bomb works, other countries would be tempted to copy the American lead, but this path must look much less attractive if there is still great uncertainty about the bomb working. This may or may not have been cogent at the time, but this is not the point; its cogency was never explored in argument until it was much too late. The time for this question was when the German threat had receded, not on the eve of the date planned for the nuclear attack on Japan.

In the light of all these points it is clear that the shift in objective of the Manhattan Project was a profound one, requiring the most serious scrutiny. In fact, as we have seen, the change was made in the almost total absence of critical scrutiny. By the time it was clear that there was no threat from a German bomb, the Project had acquired so much momentum that it could not be stopped; it had a life of its own. Of the two extreme views which are possible, therefore, the one which sees the change of objective as an *ad hoc* device used by the Project's controllers to ensure its continuation is the one closest to the truth. The next question, of course, is why this happened. A complete account cannot be given here, but the following features seem to be important.

(a) As mentioned before, the change in objective occurred smoothly over time. As intelligence gradually put paid to the fears about the German bomb, so attention shifted to an offensive weapon against Japan. This was possible because the change of objective called for no change in the efforts of the Project.

(b) The scale and pace of the Manhattan Project was breathtaking. Its enormous size, even by early 1944, meant that enormous disruption would follow its abandonment. At its handover to civilian control, the Project's payroll included 254 military officers and 1,688 enlisted men, nearly 4,000 permanent civilians and 37,800 on contract. Its thirty-seven installations were sited in nineteen states and Canada.

(c) The Project was extremely secret. Secrecy may sometimes be essential, as in the present case, but it tends to lead to bad decision making because the only people sufficiently informed to play a role in decision making all share the same fundamental views, including a commitment to the project in question. This is certainly true of the Manhattan Project. Anyone who was allowed access to sufficient information to be of some influence in determining the direction of the Project and, in particular, whether or not it should continue after the evaporation of the German threat, was either a high ranking military man, and so dedicated to the destruction of Japan and fascinated by the military potential of the new bomb; a politician likely to lose face from the cancellation of such a huge project; or a scientist employed on the Project and so committed to it and the furtherance of scientific knowledge which it represented. In these circumstances the case for the cancellation of the Project could never have been formulated cogently and defended in debate. Any argument about the future of the Project, in other words, was bound to be biased in its favour.

The politicians involved may have been expected to take a wider view than the military and scientists involved in the Project, but in this they failed. In the words of General Groves, President Truman's decision to use the new bomb was not so much saying 'yes' to the planned attack as not saying 'no'.[8]

(d) It would have been very risky for any politician to advise the cancellation of the Project. As in so many other cases, suggesting a change in the status quo would have been for the politician to open himself to attack from his opponents. Such advice could easily be interpreted as a lack of vigour in the prosecution of the war against Japan, or as admitting culpability for the decision to begin the huge Project in the first place instead of a brave attempt to remedy an error to which no blame could reasonably be attached. Again, the size of the Project worsened the problem. It is one thing for a politician to be seen as erroneously spending a million dollars or two, but quite different to be seen as wrongly disbursing the billions of the Manhattan Project.

(e) Deciding whether to continue a project already underway is, in ordinary cases, inevitably biased towards the project's continuation. Suppose it were proposed in early 1942 to develop an atomic bomb against Japan's cities, the German threat being discounted. The project would consume a huge amount of scarce resources, and so would quickly raise a crowd of objectors from people wanting to use these resources in some other way. The project is also risky, and may never succeed in producing its promised weapon. For these two reasons the case for the project is sure to receive a very close scrutiny; the case against it is certain to be developed fully and forceably. But what if many of the costs have already been borne and cannot be recovered and if much of the risk has been eliminated by research, as was the case when it was proposed to change the objective of the Manhattan Project from countering the German threat to producing a weapon against Japan?

The costs of the plan for the Japanese weapon are now much lower than they would have been in 1942 and many of the most serious risks had been eliminated. To this extent, continuing the Project after 1944 is a much more favourable enterprise than starting in 1942 with the express purpose of developing a bomb for use against Japan. But because most of the costs were sunk and the risk of failure had considerably diminished, there was little *motivation* for anyone to query the cogency of the case for continuing the Project. We have briefly considered some of the issues which were involved in the Project's change of objective, but none of these was seriously debated at the time because no actors in the decision could benefit from its cancellation. No party was, therefore, motivated to make out the case for the Project's termination, even if we can see, in retrospect, that the case was very strong. The debate, such as it was, about the change in objective was, therefore, inevitably biased – the case for carrying on was not properly scrutinized.

Having considered the case of the Manhattan Project, we may now ask what general lessons might be learned from it concerning the control of technology. It might be objected here that the Manhattan Project is so special a case; with wartime imposing enormous pressures on the pace and quality of decision making, the need for secrecy, the virtual absence of monetary constraints, the problems of gathering information about the need for the Project in wartime and so on, that little of general importance might be expected of it. I incline, nevertheless, to the opposite view. If we wish to learn about the problems of controlling technology we should look where these arise in the most pressing and urgent way. If our interest lies in the effect of restricted information on decision making, we should look at decisions made in the highest secrecy; if interested in the inertia acquired by technology, we should consider projects of great scale and urgency. All in all, therefore, there is much to be learned from the Manhattan Project where extremes of all kinds are found together.

The first lesson is that a change in the objective of a technological project needs to be scrutinized very seriously. It may represent the exploitation of some freshly discovered technological opportunity, but it may be an *ad hoc* move by proponents of the project who want to continue with it come what may and are willing to invent suitable objectives to ensure this.

The second lesson is that secrecy can seriously impair the quality of decision making about a technological project. It can happen that all those involved in the project's direction are committed to its success; potential opponents being denied access to the information necessary to make out their case. The project is then guided by a group who are all in agreement about the general direction it should take and shielded from ever hearing arguments against the project. An example even more spectacular than the Manhattan Project is British decision making on atomic affairs in the immediate post-war period, where extreme secrecy led to decisions of the greatest importance being made by a handful of men all committed to developing the new technology. We may refer to this feature of decision making as the *cabal effect*. This is recognized in Council for Science and Society (1976) where talking of the implementation of a technology it is noted that:

> At this stage, more than at any other in a technical development, the opinions of experts must be capable of effective and independent expression. All too often, however, the experts with relevant competence are committed to the project, having spent their time hammering out the very design that is now under consideration. This is partly the case in a superstar technological project, such as Concorde, whose conception and development for a long time pre-empted the services of a high proportion of the experts in a major industry in several countries.
>
> The task of providing effective monitoring power at this stage is thus more urgent and more difficult. A deliberate effort must be made to maintain a corps of experts who are not committed to the project, and they must be given sufficient incentive to subject the whole scheme to critical appraisal for its wider, non-commercial, costs and benefits. To counter the interests supporting the project, the monitors are bound to attach themselves to alternative power groups – perhaps the Treasury opposing wasteful government expenditure, or 'public interest' organizations fearful of dangerous side-effects, or those whose personal interests are actively threatened by the innovation as proposed.
>
> At this stage the monitoring process no longer lies in the realm of hypothesis and intellectual debate: it has moved into the political arena. It therefore partly takes the form of a trial of strength between power groups. The experts are caught up in an *adversary process.* [9]

The final lesson is that there is an inherent tendency to continue with a technological project, which increases as more of its costs are sunk, and as the uncertainties orginally surrounding it are resolved. This is not the result of a conspiracy on the part of the project's proponents, but arises when such a large part of the project's costs are sunk and so many uncertainties settled, that no group involved in the project's direction is motivated to argue for its abandonment. This does not amount to some kind of technological determinism, because it is always within human power to cancel the project, but it does imply that any argument as to whether the project should continue is biased in its favour. The case for cancellation will not be pressed with the same force as the case for continuing. We may call this effect the *conservation of technology*.

References

1 Hinton (1977), p. 11.
2 Braun (1980), p. 23.
3 Roskill (1971), pp. 4–5.
4 Roskill (1971), p. 8.
5 From A. Smith (1965).
6 From A. Smith (1965).
7 From R. Junk (1960), p. 171.
8 From R. Junk (1960), p. 190.
9 Council for Science and Society (1976), p. 34.

Bibliography

E. Braun (1980), *Government Policies for the Stimulation of Technological Innovation*, Working Paper 80–10, International Institute of Applied Systems Analysis, Laxenburg, Austria.

Council for Science and Society (1976), *Superstar Technologies,* Barry Rose.

M. Gowing (1964), *Britain and Atomic Energy 1939-1945,* Macmillan. (1974), *Independence and Deterrence Vol 1-Policy Making*, Macmillan.

S. Groueff (1967), *Manhattan Project: the untold story of the making of the bomb*, Little, Brown & Co.

L. Groves (1962), *Now It Can Be Told: the story of the Manhatten Project*, Harper Bros.

R. Hewlett and O. Anderson (1962), *A History of the US Atomic Energy Commission Vol 1, The New World*, Pennsylvania State University Press.

Hinton (1977), 'The Birth of the Breeder', in J. Forrest (ed.), *The Breeder Reactor*, Scottish Academic Press.

R. Junk (1960), *Brighter Than a Thousand Suns*, Penguin.

G. Leitch (1978), *Report of the Advisory Committee on Trunk Road Assessment*, HMSO.

Roskill (1971), *Report of the Royal Commission on the Third London Airport*, HMSO.

A. Smith (1965), *A Peril and A Hope*, University of Chicago Press.

J. Wilson (ed.) (1975), 'All In Our Time: The Reminiscences of 12 Nuclear Pioneers', *Bulletin of the Atomic Scientists.*

9. KEEPING CONTROL

It is now time to consider what can be done to counter the various obstacles to the control of technology which earlier Chapters have identified. In each case the problem is to be overcome in the same way – *by the buying of flexibility, controllability, corrigibility or insensitivity of performance to error*, although the form in which it is purchased will differ from case to case. The following is a guide to how the problem of enhancing the control over a technology can be approached, although the application of these ideas to particular cases obviously calls for a far more detailed appreciation than can be given here.

Entrenchment

Entrenchment is the resistance to control which arises from the adjustment of surrounding technologies to one which is developing, so that changing the latter eventually involves widespread changes to all sorts of other technologies. It is a particular problem for highly valuable, low variety systems where the operation of the system depends heavily on one technology and where its failure is very costly. It is difficult to say much at a general level about ways of improving control over an entrenched technology, so we may consider the example used before, motor-car transport. The transport system of any economy is extremely valuable; the cost of its failure is enormous. In all Western economies, however, the motor car dominates the system and surrounding technologies have adjusted to cheap and plentiful motor-car transport. Oil refineries produce the petrol required by cars, and the chemical industry has accustomed itself to using some of the remaining fractions as feedstock. In turn, the petrol producers have come to rely on lead additives in manufacturing a product of sufficient quality to satisfy the high compression ratio engines with which most cars are fitted. The scale of other forms of transport, particularly buses and trains, has adjusted to the dominance of the private car. There are, of course, social as well as these technological adjustments, as witnessed by the growth of satellite villages and outer suburbs.

All of these adjustments are difficult, expensive and slow to change, making motor-car technology very resistant to any kind of serious control. Not only does a change in the scale of motor transport require serious changes in oil refining, the chemical industry, and the geography of towns, it can only be achieved if it is accompanied by large proportional increases in bus and train transport. A reduction of a half in motor-car transport might be accommodated by an increase of 150 per cent in bus and train transport. For this reason any such shift in the means of transport must be slow and very costly, requiring very large capital investment. This does not matter providing the existing system functions smoothly and receives no surprise shocks. But shocks there are certain to be, one possibility, the discovery that lead from petrol exhausts has serious health effects on children, having been discussed in Chapter 3. Other shocks might concern the cost and availability of crude oil. What sort of action might make the performance of the transport system less sensitive to such shocks; how, in other words, can its flexibility or ease of control or corrigibility be enhanced?

(a) One policy discussed earlier is the lowering of car engines' compression ratios over a suitable period of time. The higher an engine's compression ratio, the fussier it is about the fuel which can be used without risking engine damage, so the effect of this change would be to widen the range of fuels which the maintenance of private car transport demands. Thus if lead has to be removed from petrol the adjustments necessary to engine design and oil refining would be much less than at present. Even more seriously, if other liquid fuels have to be developed to replace diminishing supplies of petrol from crude oil, the demands placed on them will be reduced so that their introduction will be easier. Such a policy would cost money, and this should be no surprise as we should expect to pay for extra control or flexibility. The cost is in the reduced efficiency of engines with lower compression ratio, calling for either a loss of performance or increased petrol consumption. Whether the cost is worth the extra control, or whether control can be purchased more cheaply in other ways is, of course, a matter of detailed analysis which cannot be undertaken here.

(b) A second way of enhancing control is to reduce the transport system's dependence on the private car, by encouraging a gradual shift to train and bus. It is then possible for these to accommodate a further drop in car transport in response to some unexpected event, such as partial failure of oil supplies, much more quickly and cheaply than at present. The cost of such a shift might be high because the car is the favoured form of transport, but the increase in flexibility of the transport system might be considered worth the expense.

(c) A further improvement in flexibility could be achieved if the above change was accompanied by a diversification of fuels used in transport. At present the transport system is heavily dependent on oil, which provides petrol and diesel fuel. Electrification of railways and the development of electric cars would enable coal and nuclear power to supply energy for transport. The whole transport system would then become much more resistant to changes in the price and availability of oil.

(d) Knowledge is a great enhancer of control, because it reduces the time required to apply controls. R & D on liquid fuels which can substitute for oil-based petrol and diesel fuel would enable any substitution which might be necessary in the future to be made more quickly. This could be done in the time taken to build the necessary plant and acquire its feedstock, without the long period needed for developing the process.

The response time could be reduced even further if the new liquid fuels were manufactured in large quantities, enabling an appreciable substitution to be made for oil-based fuels. This would be much more expensive than simply undertaking the R & D required, but the reduction in time taken to respond to some unexpected problem in oil supplies, or problem of pollution from existing fuels, might be regarded as worth the cost. This approach can be seen as a rival to (c) above, because the use of the new liquid fuels would reduce the transport system's dependence on oil. There are, therefore, various ways to enhance the flexibility of the system in this respect, but a critical comparison cannot be undertaken here.

(e) Any major shift in the transport system away from the use of petrol and diesel fuel will make the traditional oil-based feedstocks of the chemical industry less attractive. R & D on alternative feedstocks would, therefore, reduce the time and expense of the chemical industry's adjustment to such a shift. As before, the full scale development of plant using the new feedstocks would further reduce the response time of the chemical industry, but, of course, at greatly increased cost.

Competition

When competition involves developing technologies with long lead times we have seen that a disastrous pattern of decision making can occur where each party develops the technology as a hedge against anyone else developing it, so that all the rivals finish up using the technology, even if it proves to have large social costs. Competition therefore makes the control of the technology very difficult; each side seeing itself as being forced to adopt the technology whatever its social cost, because the cost of not having it when a rival has it is even higher. The following are some of the ways in which this problem can be mitigated.

(a) The simplest way is agreement not to compete in this way, for example, by no party developing the technology in question, or by limiting its rate of development so that its social consequences may be assessed. In practice, many arms control agreements are of the first kind, and counterparts in civil technology undoubtedly exist, but are harder to discover, concerning as they do, cartel arrangements and agreements to limit free competition. Such control has a cost in the foregone benefits from the technology, and in the risk from the agreement being broken, and this may be weighed against the enhancement of control.

(b) A second way of improving control in a competitive environment is to prepare for the worst that might happen, i.e. the development of some technology by a rival, but not by going to the extreme length of developing the technology oneself. Preparation may take the form of R & D which means that the technology can be quickly adopted once a competitor is known to have it. As before, R & D reduces response time, and so increases controllability. York's argument about the Superbomb was mentioned briefly in discussing MIRV, and he was making just this point. American fears that the Soviet Union would test an H-bomb first led to the American Superbomb programme, so that both sides eventually finished with an H-bomb when neither wanted it. York believes that the United States could have afforded to postpone starting the bomb programme until the Soviet Union had tested its own bomb because, with appropriate R & D, the American bomb would take only 2-3 years to develop, during which time her existing nuclear arsenal would have been an effective enough deterrent. Until such time as the first Soviet H-bomb test, the United States should have attempted to negotiate a mutual ban on weapons of this sort.

(c) A third way of mitigating the problems brought about by competition is by selective protection. The aim here is to reduce the social cost of the technology's introduction by protecting certain areas of the economy. If the United Kingdom is forced to develop microelectronics technology in those industries and services which compete in an international market, the unemployment which may result from the technology might be reduced by preventing its spread to industries and services where there is no competition, such as central and local government, insurance and banking and so forth.

The Hedging Circle

Consider the example of rival energy futures in Chapter 5. There are two forecasts of energy demand; a high one based on a projection of historical relationships and a low one from a detailed study of the efficiency of energy use which might be achieved in various sectors of the economy. Accepting the low forecast involves the risk that the expected improvements in efficient energy use will not happen, so that demand for energy will exceed supply. This will take many years to remedy and for this time energy supplies will restrict economic growth. Proponents of the high energy future argue that this risk is too great, and that the liberal assumptions in the high energy forecast must be used to avoid the danger of undersupply. In this way future energy provision is generous, so that there is an abundance of cheap energy which destroys any incentive for users to invest in ways which make its use efficient. The improvements in efficiency cited by the low energy forecasters therefore do not happen, and the decision to increase energy supplies appears to be vindicated. In this way, avoiding the risk of undersupply entails that very little is learned about efficient energy use, because the incentive for it vanishes. Because little is learned, the shift to efficient use and low energy consumption always remains a risk too great to take and so energy consumption continues to rise. A low energy future is thereby made unattainable. What can be done to mitigate this situation?

The essential action is not to try to determine which is the best forecast, for this is beyond our present abilities. Since any forecast is very likely to be wrong by an appreciable margin, *what is called for is action which reduces the cost of its error*; in other words action which increases the controllability, flexibility and corrigibility of the energy system and decisions about it. Hedging by making generous energy provision in the future is thought to be necessary because the cost of error, i.e. of undersupply, is reckoned to be very high. But there are all kinds of things which can be done to reduce the cost of error, so that hedging in this way is no longer needed. Future energy demand need not then be reckoned so generously, because if it does prove higher than anticipated then the system can adjust quickly to provide more energy, the cost of undersupply during the adjustment period being low enough to risk. In this way the low energy future, frustrated by conventional hedging, may be aimed for. If the critics prove to be right and consumers cannot achieve efficient use of energy, then forecasts built on this assumption will be too low, and energy supplies based on these forecasts will be too low. But if the system's ability to respond quickly to undersupply has been purchased, this need not be a disaster.

The problem of responding to undersupply is one of lead time. It is generally thought that the lead time to produce more energy in whatever forms is required is so long that the damage done before undersupply can be corrected is very large. This is true if the most economically efficient plant is required. Taking advantage of the economies of scale thought to exist in electricity generation, for example, requires building plant over seven to ten years, but it is quite possible to meet a shortfall in electricity supply by building several small stations, or extending existing ones, with a lead time of around three years, especially if planning has been undertaken in advance. Alternatively, old stations may be brought out of retirement until larger modern ones have been built. All this adds to the cost of electricity, but the risk of bearing this cost may be worthwhile to avoid the almost inevitable overcapacity which results from conventional hedging. Open-cast coal reserves are another source of flexibility because they can be exploited much more rapidly than deep-mined coal in a response to an unexpected shortage of this primary fuel. Leaving reserves in the ground means postponing exploiting the most easily won coal, and so is a cost, but it is the cost of increasing the flexibility of the energy system. In the same way, particularly accessible reserves of gas and oil could be left underground awaiting any unexpected failure to meet demand for these fuels. Other ways of enhancing flexibility will emerge from the case study which follows.

What is needed as a replacement of the sterile debate about what forecasts are to be believed, is a study of the host of ways in which the energy system's adaptation to an undersupply of energy can be enhanced, and what they will cost.

Lead Time and Unit Size

Flexibility or ease of control varies inversely with lead time and unit size of a technology, so that flexibility may be enhanced by adopting technologies with a low lead time and/or small unit size. As discussed in

Chapters 6 and 7, this will usually be at the cost of losing economies of scale, but flexibility must be paid for.

Dogmatism

Dogmatism is a danger where those who monitor a decision to develop a particular technology have a vested interest in its continuation. They can then exploit a whole battery of devices for evading criticism so that the cherished project may continue. How can those monitoring the decision be forced to be less cavalier with criticism of their project? We have considered three factors which tend to heighten the risk of dogmatism; secrecy; the cabal effect where the decision makers are all dedicated to the project and the conservation effect where projects which have absorbed considerable resources become increasingly difficult to stop. The lesson, of course, is the avoidance of each of these. Monitoring a technology should be performed in the minimum of secrecy, those responsible for monitoring should have a range of backgrounds and some should not be closely associated with the technology, and scrutiny of on-going projects should be particularly severe. All this would seem to require some kind of adversary method of monitoring, of the sort favoured by the Council for Science and Society:

> The only known way of countering the fallibility of one individual is to subject his opinions and decisions to the criticism of others. This is the principle of pluralism by which we run a democratic society. It is also the principle by which a body of reliable scientific knowledge was won for mankind. *The control of advanced technical projects on behalf of society must depend on the same principle as does science, and therefore requires the strengthening of critical scrutiny inside and outside the corporate agencies of technical change*. This is what we call monitoring technology, but we mean much more than simple surveillance and retrospective sanctions against proved transgressors. Rather we mean a range of institutions and social techniques enabling the critical scrutiny of corporate decisions and actions, by and on behalf of competent and concerned opinion at every level.[1]

This should be clearer after the digestion of Part 2.

Case Study – Energy R & D

The ideas of Chapter 2 are now to be applied to the problem of what R & D programmes in the field of energy should be underway now, and in the future, in the United Kingdom. The fruit of such programmes, new energy technologies, is long in the bearing, several decades of R & D preceeding many energy technologies. As observed in Chapter 2 this long lead time makes nonsense of any attempt to apply Bayesian decision theory because the costs and benefits of an R & D programme stretching over decades just cannot be estimated. Forecasting with sufficient precision and accuracy over such time scales is difficult enough anywhere, but impossibly so in the field of energy.

The briefest acquaintance with the history of British energy policy is enough to confirm this scepticism, for it shows the remarkable difference in the speed of technological developments and the rate at which fundamental policy changes occur. At first, coal is seen as providing energy indefinitely so that there is no need to develop other sources and the technologies associated with them, nor any need to worry about a thing called energy policy. Next, coal is seen as being replaced more or less completely by cheap imported oil, with the disappearance of gas, since this was produced from coal. After this, doubts about the availability and political susceptibility of oil began to highlight the need for nuclear energy and a future based on coal and atomic power seemed inevitable. Around this time the discoveries in the North Sea of oil and natural gas led to further changes in policy, postponing the time when nuclear power and coal would share the honour of providing the bulk of the country's energy. Nuclear power however, proved far harder to develop than anticipated, so that at the present it provides a far smaller fraction of British primary energy than anyone anticipated twenty years ago. The future balance between coal and nuclear has, therefore, shifted more and more to coal.

All these dramatic changes in the view of the future of the country's energy system are compressed into the last thirty years or so. It is salutary to compare this with the lead time for the development of nuclear power. The United Kingdom has failed to achieve a steady programme of nuclear power stations, largely through technical problems, despite thirty years or so of continuous development. If the first commercial breeder reactor is ordered now it will not be commissioned for at least fifteen years, and it will be ten to twenty years after that before reactors of this type are able to make a significant contribution to energy supplies. And yet the first plans for breeder reactors were discussed in the late 1940s, giving a lead time from inception to significance of the order of sixty years.

Decisions about what R & D programmes to invest in now cannot hope to receive assistance from Bayesian decision theory, the uncertainties are just too huge. These are, of course, decisions under ignorance. The problem is put succinctly in Marshall (1976), which forms the principal part of our case study:

> Unfortunately, new technologies take a long time to develop and apply; the period from the inception of an initial R & D programme to successful commercial application tends to be measured in decades rather than years. It is vital, therefore, to identify those technologies which seem most likely to be important in the future and to consider how best we can ensure that they will be available to us when required. In addition, because the future is so uncertain, we must seek to keep our options open over a wide range of technologies, any of which might become important under particular circumstances ... Our ultimate objective will be to formulate a strategy sufficiently robust to cope with the immense uncertainty of the future.[2]

This calls for scenario analysis:

> The United Kingdom's future technology requirements in energy are inextricably linked to future events, most of which cannot be forecast with any degree of certainty; yet one of the main purposes of an energy R & D strategy must be to ensure that future policy makers are equipped with appropriate energy technologies to implement their policies whatever shape the energy economy takes. Therefore, it is clear that we need to adopt an analytical approach which assesses technology needs over a spectrum of possible futures, and not just on some 'central' view. Of the many techniques available, we have adopted an approach in which the contribution of each technology is assessed over a set of narrowly defined 'snap-shot' views of the future – these are termed 'scenarios'.[3]

The objective of the study, to find strategies which can 'cope with the immense uncertainty of the future', is not very precise because what is it to 'cope'? Coping seems to be the provision of sufficient energy in appropriate forms not to place a constraint on the growth of the British economy. This is clear from Marshall's approach where he calculates the energy needed to sustain the British economy from the assumptions in the scenario about world and British economic growth.

A scenario may be thought of as a hypothetical forecast of form. If $D_1 \ldots D_k$ and $S_1 \ldots S_m$ then $O_1 \ldots O_n$, where $D_1 \ldots D_k$ are decision variables, $S_1 \ldots S_m$ state variables and $O_1 \ldots O_n$ outcome variables. A set of scenarios is a set of such hypothetical forecasts with different values for the decision and state variables. In a proper use of scenarios, the analyst is concerned with the sensitivity of outcome values to change in decision values. For convenience, the outcome variables may be represented by a single variable having just two values, acceptable and unacceptable. Letting D_i^j be the jth value of the decision variable D_i, the outcome is more sensitive to D_i^j than to D_i^h if there are fewer satisfactory outcomes given $D_i = D_i^j$

than satisfactory outcomes given $D_i = D_i^h$. If the aim of the decision maker is to achieve an acceptable outcome, then setting D_i to D_i^h gives him at least as great a chance of this than setting it to D_i^j. The point of the exercise, of course, is that future values of state variables are not known, and cannot be ascribed a probability distribution, so that choosing D_i^h instead of D_i^j is making a decision which is more robust to whatever events come to pass.

The exposition of Marshall's scenario analysis does not follow the pattern discussed above and, indeed, is somewhat confusing (e.g. no clear division between state and decision variables is made). Nevertheless, it is easy to restructure the analysis along the lines suggested earlier. The problem is to choose a set of R & D programmes which will be robust in the sense of allowing the future British energy system to be in an acceptable state. For the sake of simplicity, we may think of an acceptable state as one giving a balance between primary energy, energy carriers and useful energy.

Primary energy available to the United Kingdom is in the form of coal, natural gas, oil, uranium and natural forces such as the wind and tides. To be usable, however, primary energy must be processed from its raw state and delivered to energy consumers. This generally involves conversion of the energy to a different state, as for example when coal is burned in a power station to produce electricity, or where crude oil is refined to yield petrol, and these are called energy carriers. Energy is delivered in the form of energy carriers, such as electricity and petrol, and then these are used to fulfil some human function such as keeping warm or making some product. The energy used in this way is called useful energy and is a fraction of the energy delivered by carriers, the rest being lost. Using energy in this way, say to transport goods, requires that it is delivered in an appropriate form. Most existing transport technologies demand liquid fuel as a carrier, and these fuels are only available because there are technologies which can convert primary energy, in the form of crude oil, to liquids like petrol and diesel. These technologies are, in turn, only possible because others exist for winning and transporting crude oil. The technologies we possess therefore ensure a balance between the useful energy which is demanded, energy carriers and primary energy. Any economy has a particular spectrum of demand for useful energy delivered by various carriers, obtained from a variety of primary energy sources. If the technologies required to ensure a balance between useful energy, energy carriers and primary energy do not exist, then that economy cannot function.

The problem therefore, is to invest in those R & D programmes which will enable technological choices to be made in the future so that useful energy, energy carriers and primary energy can be balanced and the economy sustained over a wide range of possible futures. We cannot know how the energy system will function in the future, but what we can try to do is provide future decision makers with the know-how to ensure that their system is balanced over a wide range of possible futures. Each scenario can, therefore, be seen as having the following form, the period of interest for Marshall being 1975–2025;

> If R & D decision $R_1 \ldots R_n$ are made at times $t_1 \ldots t_k$ and if $S_1 \ldots S_m$, then there will (will not) be a balanced energy system from 1975–2025.

$S_1 \ldots S_m$ are the scenarios' state variables, or state of the world variables. The object of the exercise is to find a set of R & D decisions which lead to a balanced energy system over a wide range of values of the state variables, i.e. a set of R & D decisions to which the outcome of a balanced energy system is highly insenstivie. Marshall presents seven scenarios, which are outlined in Table 9.1.

The economic factors in the scenario are first used to calculate useful energy requirements in the three sectors, building, industry and transport. By means of a simple energy model requirements for the heat supplied by various energy carriers (solid, liquid and gaseous fuels, electricity, district and direct heat) are calculated. The primary energy in the form of coal, oil, natural gas, nuclear power and alternatives needed to meet this requirement for carriers is then calculated. Thus calculated, there is obviously a balance between primary energy, energy carriers and useful energy. To actually achieve this balance within the scenario, however, requires new technologies to be available in the three areas; primary energy exploration and provision, energy conversion and distribution and energy utilization. The technologies which are required for the scenario's energy system to be balanced are therefore identified and the timing of their introduction estimated.

For example, Scenario 5 describes a self-sufficient energy system where primary energy is nuclear and coal. Today, however, oil is the chief provider of liquid carriers for transport, so that the energy system of the scenario can be balanced only if there are technologies for the conversion of coal to liquid fuels or for the electrification of transport. The R & D for this must be undertaken soon, because of the long lead times of energy technologies, if the option of a self-sufficient energy system is to remain open in the future. When all the scenarios have been treated in this way, the sensitivity of obtaining a balanced energy system to the

*Table 9.1 – Energy Scenarios**

	State-of-World Assumptions						
State-of-World Variables	0 *Trends-Continued Scenario*	1 *Low-Growth Scenario*	2 *Limit-on-Nuclear Scenario*	3 *High-Energy-Cost Scenario*	4 *Price-Transition Scenario*	5 *Self-Sufficiency Scenario*	6 *High-Growth Scenario*
(1) UK Economic Growth (GDP average annual)	Central view	Depressed	Reduced	Reduced Slightly	Central view to 1990, then reduced	Close to Central view	Buoyant
1975-1990	3%	2%	2½%	2½%	3%	3%	4½%
1990-2000	3%	1½%	2½%	2½%	2%	3%	4%
2000-2010	2½%	1%	2%	2½%	2%	2½%	3½%
2010-2025	2%	½%	1½%	2%	2%	1½%	3%
(2) World Economic Growth	Central view, 1–1½% above UK growth	Depressed ½–1% above UK growth	Low, ½–1% above UK growth	Reduced ½–1% above UK growth	Central view to 1990 then reduced	Central view, 1–1½% above UK growth	Buoyant 1½–2% above UK growth
(3) World prices of Energy Raw materials	Rising slowly until 1990s, then faster, exceeding cost of syncrude soon after 2000	Slightly below Scenario 0 levels	Rising rapidly from early 1980s onwards	Sharply higher doubling immediately, stabilizing until 2000, then as in Scenario 0	Stable until 1990, then jumping well above Scenario 3	Similar to Scenario 0 levels	Rising somewhat faster than Scenario 0
(4) UK Internal Energy Prices (in relation to world prices)	Present relationship	High	Slightly higher than present relationship	Present relationship	Present relationship until 1990, then high	High	Present relationship

Table 9.1 (continued)

(5) Energy Conservation Effects	Present expectations	High, but delayed by low economic activity	Very high, well above present expectations	Very high, well above present expectations	Present expectations to 1990, then very high	Higher than on Scenario 0	Present expectations but accelerated by high economic activity
(6) Premium on Self-Sufficiency	Present policy	Present policy	Present policy	Present policy	Present policy	Very high	Present policy
(7) Nuclear Plant Building Programme	Up to a maximum of 50–60GW(e) installed in year 2000	Up to a maximum 40–50GW(e) installed in year 2000	Programme halted after SGHWRs commissioned in mid-1980s	Up to a maximum of 50–60GW(e) installed in year 2000	Up to a maximum of 50–60GW(e) installed in year 2000	Up to a maximum of 90–100GW(e) installed in year 2000	Up to a maximum of 100-110GW(e) installed in year 2000
(8) Prospects of Energy from Alternative Sources	Present expectations	Below present expectations	Very high, well above present expectations	Above present expectations	Present expectations	Present expectations	Present expectations
(9) Coal Production	Present plans and expectations	Slightly above present plans and expectations	Well above present plans and expectations	Above present plans and expectations	Present plans to 1990, then rising above present expectations	Above present plans and expectations	Present plans and expectations

**Source*: Marshall (1976), p. 19.

Notes: (1) The same demographic assumptions are common to all scenarios, namely: total population would slowly rise but not exceed the 1975 level by more than 10 per cent in 2025; however, the total productive workforce is not expected to increase significantly over this period.

(2) Structural shifts within the economy are assumed to have only a second order effect on energy consumption. Industrial output has been assumed to rise at a rate of 0.25 percentage points above the GDP rate on all scenarios.

*Table 9.2 – Research Priorities and Main Implementation Options**

Technology	*Overall Importance of the Technology*	*Ongoing Phase of R, D & D*	*Priority For UK Involvement in Ongoing Phase*	*Appropriate Option for Implementation of Ongoing Phase*
(1)	(2)	(3)	(4)	(5)
PRIMARY ENERGY TECHNOLOGIES				
Coal Production				
Mining technologies	x x x x x	R, D & D	High	(a)
In-Situ Conversion	x	Research	Low	(e)
Oil and Natural Gas Production				
Continental Shelf Technologies	x x x x x	D & D	High	(a) & (c)
Enhanced Recovery of Oil	x x x x	R, D & D	Medium	(c)
Stimulation of Gas Reservoirs	x x	R, D & D	Low	(c)
Deepwater Oil and Gas	x x x x	R & D	High	(a) & (c)
Nuclear Energy				
Uranium Supply	x x x x x	Exploration	Medium	(c)
Fuel Processing & Reprocessing	x x x x x	R, D & D	High	(a)
Thermal Reactors	x x x x x		High	(a)
Fast Reactors	x x x x x	Demonstration	High	(a)
Radioactive Waste Management	x x x x x	R, D & D	High	(a)
Nuclear Safety	x x x x x	R, D & D	High	(a)
Nuclear Process Heat	x x	Development	Low	(d)
Alternative Energy Sources				
Fusion Power	x x x	Research	Medium	(b)
Geothermal Heat	x	Survey	Medium	(d)
Solar Heat	x x	D & D	Medium	(a)
Solar-Photovoltaic	x	Research	Low	(e)
Solar-Biomass	x	Research	Low	(e)
Tidal Power	x x	Assessment	Medium	(b)
Wave Power	x x x	R & D	High	(a)
Wind Power	x	Assessment	Low	(d)
Oil Shales	x	Survey	Low	(d)
Wastes	x x	D & D	Low	(d)
ENERGY CONVERSION & DISTRIBUTION				
Coal Conversion				
Coal as a Power Station Fuel	x x x x	R, D & D	High	(a)
SNG from Coal	x x x x	D & D	Medium	(b)
Synthetic Hydrocarbon Liquids from coal	x x	R, D & D	Low	(d)

Table 9.2 (continued)

Metallurgical Cokes from Coal	x x x x	R, D & D	High	(a)
Electricity Supply				
Electricity Generating Plant	x x x x x	R, D & D	High	(a)
Transmission and Distribution	x x x x x	R, D & D	High	(a)
Electrical Bulk Storage	x x x	R & D	Low	(d)
Combined Heat and Power Plant	x x x	(Commercial)	–	–
Gas Supply				
SNG from Oil	x x x	R, D & D	High	(a)
Transmission, Storage and Distribution	x x x x x	R, D & D	High	(a)
Other Energy Carriers				
District Heat	x x	(Commercial)	–	–
Hydrogen	x	Research	Low	(e)
ENERGY UTILISATION TECHNOLOGIES				
Utilisation of Fuels				
Coal as a Domestic and Industrial Fuel	x x x x x	D & D	High	(a)
Electricity Utilisation Technologies	x x x x x	R, D & D	High	(a)
Electric Traction	x x	R, D & D	Medium	(b)
Gas Utilisation Technologies	x x x x x	D & D	High	(a)
Heat Pumps	x x x	D & D	Medium	(b)
Alternative Transport Fuels	x	R & D	Low	(e)
Energy Conservation Techniques				
Conservation in Buildings	x x x x x	R, D & D	High	(a)
Conservation in Industry	x x x x x	R, D & D	High	(a)
Conservation in Transport	x x x x x	R, D & D	High	(a)
SUPPORTING RESEARCH STUDIES				
Basic Research	x x x x	Research	High	(a)
Energy Systems Studies	x x x x	Research	High	(a)
Environmental Studies	x x x x x	R & D	High	(a)

Source: Marshall (1976), p. 27.

Notes: (1) In column 3 R, D & D = research, development and demonstration; R & D = research and development; D & D = development and demonstration.

(2) In column 5 (a) = take a national lead; (b) = maintain a good technical competence (c) = rely on international commercial interests; (d) = acquire and keep the status of informed buyer; (e) = await developments elsewhere.

R & D decisions is estimated. Some R & D programmes are involved in nearly all the scenarios, so that the outcome is not sensitive to them, while others figure in few scenarios, so having a greater sensitivity. Table 9.2 above gives the contributions of some of the technologies and their overall importance to ensuring a balanced British energy system.

Three areas emerge as particularly important, energy conservation, coal and nuclear energy. Conservation ensures the efficient use of energy and is therefore important on all scenarios, so that R & D in improved methods of conservation is a priority. Coal is an essential primary fuel in all the scenarios so R & D in deep mining, direct utilization and coal conversion are important, as is R & D in nuclear energy which is required in all but one of the scenarios. Shortages of uranium are expected to necessitate the introduction of the fast breeder reactor which uses this fuel more efficiently than existing thermal reactors, and this too generates a programme of R & D.[4]

The connection between the scenario analysis and decision making under ignorance may now be discussed a little more fully. Decisions about energy technologies are, of necessity, under ignorance because of the great lead times which are involved. R & D is, therefore, required to keep future options open. Engaging in the R & D programmes investigated by Marshall is essentially buying flexibility in the form of future options to implement technologies should they be needed. The aim is to maintain a balanced energy system, and there are many ways in which this balance might be achieved. Many of these involve nuclear technology so that adopting R & D programmes for this technology means that many future ways of balancing the system are possible. Opting for a nuclear energy R & D programme, therefore, places far fewer constraints on what other decisions may be made to maintain a balanced energy system than not adopting such a programme. If, on the other hand, a particular technology figures in only one scenario, then not investing in an associated R & D programme closes only a few future balanced energy systems. A limited R & D budget may mean that the programme competes with a nuclear energy programme, but the latter is clearly more important in keeping open future options. Putting the matter formally: 'nuclear energy programme' is a two valued decision variable (this of course is a simplification but the general point still holds) having values 'yes' and 'no'; there are more balanced energy systems across the scenarios which are possible given the value 'yes' than given the value 'no' hence 'yes' is a less sensitive value for 'nuclear energy programme' than 'no'. In general, insensitivity may be taken as a measure of the decision's flexibility.

Flexibility may also be seen as a property of a system, reflecting the ease with which it may be controlled. In the case considered by Marshall, for example, the system is the British energy system and control is the maintenance of balance between primary energy, energy carriers and useful energy. Error costs arise when this balance is not achieved, and can be lowered by the introduction of technologies. These technologies have, however, very long lead times so that controlled error costs would be extremely high if they were developed only when seen to be necessary. Response time may be reduced by investing in R & D programmes before the technologies are known to be required. In this way, controlled error costs are reduced so that, according to the measures proposed earlier, the effect of investing in R & D is to increase the controllability of the energy system. This is accompanied by an increase in its control costs, because R & D programmes must be paid for. The decision maker must, therefore, trade off improving his control of the energy system and increasing the control costs of the system. Investing in, say, nuclear power R & D is shown by scenario analysis to be effective in reducing error costs, while investing in a technology which figures in very few scenarios is likely to be much less effective. Scenario analysis, therefore, gives guidance as to the increase in the controllability of the system purchased in various ways.

The third way of viewing decisions under ignorance is to favour options which are easily corrected. If the choice of such an option is later revealed to be wrong then the error will not impose great costs and it may be remedied quickly. Suppose that R & D decisions are based on a single forecast of Marshall's state variables, which are expected to have the values in Scenario 5. On this scenario, a choice has been made in the future to achieve self-sufficiency in energy, so nuclear power is very important and its associated R & D would be given priority. If the forecast of the state values is later shown to be wrong, the R & D decisions made on the basis of it are also shown to have been made for the wrong reasons, but nuclear energy may still be required, if on a reduced scale, so that investment in its R & D may not have been wasted. In fact, the scenarios show that this is very likely as nuclear power appears in all but one of them. Thus investing in nuclear R & D is a highly corrigible option; if the forecasts which originally provided the reason for the investment turn out to be wrong, it is likely to be required in any case. With R & D for electric traction, however, the reverse is true. If the forecast that the future will be close to Scenario 5 is made, it is important to find substitutes for oil-based liquid fuels in transport, and electric traction is one option for this. It is, however, required on no other scenario. If the forecast proves inaccurate therefore, any R & D investment in electric traction will be lost. In other words, the investment is hard to correct, and is therefore not to be highly favoured. The same arguments show that the performance of the energy system over the time in question is less sensitive to forecast errors if nuclear R & D is chosen than if R & D on electric traction is chosen; this, of course, being the fourth way in which to view favoured options in decision making under ignorance.

Whatever is thought of the details of Marshall's work, and he clearly states the provisional nature of his findings,[5] it is clear that the technique of scenario analysis is a valuable one in assessing flexibility in decision making under ignorance.

In conclusion we may return to the question posed towards the end of Chapter 3, can the entrenchment of a technology be foreseen, and what can be done to prevent it? Yet another way of looking at Marshall's problem is that he realizes that if R & D decisions are based on narrow forecasts of the future, the technologies which will result will inevitably become deeply entrenched. If, say, no R & D is done on nuclear technologies, the economy must come to rely on coal and oil as primary fuels and other technologies will adjust to this. To take just three examples, there will be no incentive for the development of electric traction, or for liquid fuels for transport which are not based on oil, or for finding feedstocks for the chemical industry other than oil. Thus, the functioning of the energy system will come to rely on technologies which exploit the two primary fuels coal and oil, so that any unexpected shock to the supply and cost of these will be extremely hard to control, and so expensive. This may be countered by R & D on nuclear power and on the three technologies above and others like them which enable nuclear energy to be exploited. Such a programme of R & D would enable a shift to be made to nuclear energy from coal and oil, even if it is originally decided that there will be no nuclear power, and so the programme diminishes the entrenchment of coal and oil technologies. Whether the cost of R & D is justified by the diminution in entrenchment is, of course, a matter which cannot be handled by analysis.

References

1 Council for Science and Society (1976), p. 27.
2 Marshall (1976), p. 9.
3 Marshall (1976), p. 16.
4 See Chapter 6.
5 For an expansion of his work see Department of Energy (1979).

Bibliography

Council for Science and Society (1976), *Superstar Technologies*, Barry Rose.

Department of Energy (1979), *Energy Technologies for the UK*, Department of Energy, Energy Paper 39, HMSO.

W. Marshall (1976), *Energy Research and Development in the UK*, Department of Energy, Energy Paper 11, HMSO.

PART 2

VALUES AND MONITORING

10. THE LOGIC OF MONITORING

Part 1 has been concerned with the need to retain the ability to alter decisions about technology if they are discovered to have been wrong, and the obstacles which stand in the way of this. The emphasis has been placed on the response to error, not on its discovery, and the case studies have been selected accordingly. The discovery of error in the decisions which figure in the case studies is particularly straightforward. In the Manhattan Project, for example, the original decision to build an atomic bomb was brought into question when it was discovered that there was no German bomb project to counter. The hypothetical breeder reactor programme was shown to have been wrong by the fact that the reactors produced no electricity. In the MIRV example, the decisions to develop and deploy the technology were shown to be in error by the discovery that the feared Soviet developments which MIRV was supposed to counter had not been realized. Not all monitoring of decisions about technology is, however, this clear cut, and Part 2 will explore the nature of monitoring, and the problems it faces, in more depth. If decisions about technology are to be made in a more rational way than at present they need to be properly monitored, and so it is necessary to state as clearly as possible how they ought to be monitored.

The starting place, as might be expected, is the theory of decision making under ignorance of Chapter 2. This calls for decisions to be falsifiable by facts. This leads to a philosophical puzzle which, perhaps surprisingly, shows the way to a more general account of monitoring which can be used to analyse past decisions, to assist in formulating answers to contemporary issues of policy, and to discover the way to a fully developed view of decision making as a process to be explored in the final Chapter.

The adoption of one course of action in favour of others is a special case of the acceptance of a value judgement. It is to say that this course of action is *preferred* to the others which are open, that this is the course of action which *ought* to be taken, that it is the *best* of the possible actions; which are all just special cases of saying that one thing is preferred to others, that a certain thing ought to be done, and so on. In monitoring a decision, a search is being made for a fact which will reveal the decision to be false, and here lies the puzzle. Making a decision is a special case of adopting a value judgement but a very large body of philosophical thought holds that facts cannot falsify value judgements. Thus, despite their intuitive appeal, the cases we have so far discussed where a decision is shown to be wrong by the discovery of some fact run counter to a dominant doctrine within philosophical ethics.

The roots of this doctrine go very deep indeed, but our present purposes call for no more than a cursory glance at them.[1] A central question of ethics is that of how any kind of value judgement can be justified. Clearly, one value judgement, once accepted, serves to justify another, as when 'Jill ought to be flogged' is justified by appeal to the general principle 'all thieves should be flogged' and the discovery that Jill is a thief. But this cannot go on for ever. The value judgement J_2 can justify the value judgement J_1 only if J_2 itself is justified, and if the only way of doing this is appealing to some third value judgement we are obviously in an infinite regress, in which case our attempt to justify J_1 has failed. Attempting to justify J_1 leads to an attempt at justifying J_2, leading, in turn, to an attempt to justify J_3 and so on *ad infinitum.*

It follows that if any value judgement is to be justified, then the regress must terminate somewhere; at least some regresses of the above kind are not infinite. This requires the existence of some value judgements which can be used to justify other value judgements, but which are themselves so transparent that they do not need justification from other value judgements. We may term these *fundamental value judgements*. If any value judgement is justified, it is because it can be shown to follow from some set of fundamental value judgements, so that these fundamental judgements are the source of any knowledge we may have of ethical principles.

Needless to say, a great deal of effort has gone into the unearthing and clarifying of these fundamental value judgements without which it seems impossible to avoid total scepticism about evaluation. Many different characterizations have been offered. A once popular account held that as we proceed along a regress in an attempt to justify a value judgement: 'The ultimate propositions at which we arrive seem not to express mere brute facts, but facts which are self-evidently necessary'.[2] In the same vein, talking of the apprehension that something is our duty:

This apprehension is immediate, in precisely the same sense in which a mathematical apprehension is immediate ... Both apprehensions are immediate in the sense that in both, insight into the nature of the subject directly leads us to recognize its possession of the predicate; and it is only stating this fact from the other side to say that in both cases the fact apprehended is self-evident.[3]

These intuitionist views have gone out of fashion, but the problem which prompted them persists and has given rise to many more sophisticated views about fundamental value judgements than those above. It is not necessary, however, for us to become entangled in the fretful quarrels between rival philosophical factions. Instead, let us explore the link between the need for fundamental value judgements, whatever final account may be given of them, and our concern with how a fact can falsify a value judgement. If a person is aware of some set of fundamental value judgements which lead him to accept other value judgements which they entail, he must know these other value judgements to be correct. They must, after all, be justified because they follow from some set of fundamental value judgements which are themselves directly justified. If this is so, then nothing can happen which shows his accepted value judgements to be erroneous so that he must change them. In particular, no facts may be discovered which show his accepted value judgements to be wrong. As everyone learns sooner or later, facts cannot be altered to fit in with the values one wishes to accept. The only way, therefore, of ensuring that no falsifying facts are discovered is to maintain that it is altogether impossible for facts to falsify value judgements. If this is possible, then value judgements which follow from fundamental value judgements cannot be regarded as justified because they would always stand in jeopardy from factual falsification. In philosophical jargon, value judgements are *autonomous*; they cannot be established by facts nor can facts show them to be false.

It can be seen now that the apparently humble claim about monitoring made earlier is really a very serious one, potentially damaging to whole schools of philosophical thought. If facts can reveal a decision to be in error, then sometimes facts can falsify value judgements. If this is possible, then even those value judgements deducible from fundamental value judgements cannot be regarded as justified because they exist in peril of falsification from some factual discovery. In this case, the whole attempt to show that justification of value judgements is possible because somewhere there are fundamental value judgements collapses. The doors are thrown wide open to the flood of scepticism.

The doctrine of autonomy is plausible as far as it applies to decisions made under certainty, risk or uncertainty (in its restricted sense). In the first case, all the relevant facts are known to the decision maker, so unless he is wantonly stupid he should decide in such a way that no facts which he becomes aware of later can falsify the decision. Similarly for risk and uncertainty, where some closed set of possible futures can be assigned a probability in one way or another, so that whichever turns out to be realized has been taken into account in the decision and cannot, therefore, show the decision to be false. If a gambler loses heavily, this does not show his decision to bet was wrong, because he was aware that this might happen, and able to assign a probability to its occurrence and so to take it into account in deciding whether to bet. Thus the concentration of decision theory on uncertainty and its refusal to recognize the even more recalcitrant category of ignorance, enables decision theory to live happily alongside views of evaluation which incorporate the doctrine of autonomy. Indeed, in most applications of Bayesian theory autonomy is tacitly assumed in the way the decision maker's values are determined. If they are, for example, determined by asking questions about lottery gambles then, providing the answers are consistent and stable, there is no room for questioning the values assigned by the decision maker. In particular, there is no search for facts which might reveal his values to be mistaken.

When the category of ignorance is recognized this happy co-existence cannot survive. A decision maker operating under ignorance is unaware of some facts which are relevant to his decision, so whatever choice is made is in jeopardy from the discovery of some unwelcome fact. There is no way in which he can take into account such discoveries at the time of the decision in the way this can be done under happier conditions of certainty, risk or uncertainty. At a purely intuitive level this must be the case. If a decision is made because it is a way of achieving some desired state Y, and if future discoveries reveal that it does not lead to Y and, in addition, that it has quite unexpectedly, a whole spectrum of consequences not wanted by the decision maker, then what option is there but to say that his original decision was wrong? Unlike the gambler, he was not able to take these possibilities into account at the time of deciding because he did not know about them. So much for intuition. What is now needed is an account of the *logic* which stands behind this intuition.

The doctrine of autonomy may be shown to be in error by some very simple counter-examples employing only elementary logic. Consider the argument 1:

You ought to wear a coat on a rainy day.
It is not the case that you ought to wear a coat today.
____
Today is not a rainy day. . . . 1

Here the evaluative premises entail a factual conclusion. Re-arranging 1 yields 2, where a factual sentence entails an evaluative conclusion:

Today is a rainy day.

---

not-(You ought to wear a coat on a rainy day and it is
not the case that you ought to wear a coat today). . . . 2

Examples like 2 provide genuine counter-examples to the doctrine of autonomy. I interpret the force of 2 in the following way. The fact that it is raining today shows that the two evaluative sentences, 'you ought to wear a coat on a rainy day' and 'it is not the case that you ought to wear a coat today' cannot be held together. Clearly, the two evaluative sentences are not contraries, since there are circumstances, for example sunny weather, when they could be both accepted. What are we to say of someone who holds the two evaluative sentences, despite the fact that it is raining today? Odd as it sounds, I see little option but to say that this person's evaluations are *factually incorrect*. He is not being logically inconsistent, but merely unaware that his evaluations have a consequence which is, *as a matter of fact*, false.

There is, of course, nothing startling in the logic of argument 2. In applying any general evaluative principle, one has to conjoin it with a factual sentence as in 3:

All men are wicked.
Fred is a man.

Fred is wicked. . . . 3

The argument has the structure N_1 and F, therefore N_2 where F is a factual sentence had N_1 and N_2 value judgements. Any such argument can be re-arranged to given an argument of the same kind as 2 by writing: F, therefore not-(N_1 and not N_2). Consider a more serious example. X accepts both the evaluations 'all Jews are bad men' and 'X is a good man'. What happens when X learns that his true parentage has been kept from him, and that he is himself a Jew? Obviously he cannot retain his values. The newly revealed *fact* about his origins conflicts with his original *evaluations*. Since we cannot change facts to suit our values, a truism we all learn sooner or later, X has no choice but to revise his evaluation of Jews or of himself. His original position was certainly not inconsistent, but has merely been shown to be *factually incorrect*.

It must be agreed, therefore, that value judgements may be falsified by facts, contrary to the doctrine of autonomy. Any example of such falsification will be of the above form, where a factual sentence entails the falsity of some conjunction (usually a pair) of value judgements. I can see no way by which a factual sentence might entail the falsity of a single value judgement. For this reason it will always be possible to make different responses to falsification. Granted that the factual sentence is accepted, one of the value judgements in the falsified conjunction must be rejected, but which one is a matter quite beyond the settlement of logic. Thus, a factual discovery can never *force* the abandonment of a single value judgement; all it can do is to make its retention dependent upon abandoning other value judgements. Thus facts cannot determine what value judgements we may hold, but merely place restrictions on what is acceptable.

It should be observed at this point that the above account of falsification depends in no way on any particular characterization of value judgements, for example, on the view that they have a truth value. The relationship required for falsification remains even when value judgements are treated as pure imperatives, having, of course, no truth value. Thus 'If the 12.55 weather forecast says it will be showery, cancel this afternoon's match' and 'don't cancel this afternoon's match' clearly entail 'the 12.55 weather forecast did not say it will be showery'. If the entailed factual sentence is found to be false, then one or both of the two imperatives must be given up. Otherwise contrary orders are given, 'cancel the match' and 'don't cancel the match'.[4]

It may be puzzling to contrast the depth of the doctrine of autonomy with the slightness of the examples which destroy it. The explanation is that although the logic used in the case against autonomy has been known for ever, cases like the ones I suggest have been relegated to the philosophical curiosity shop because their interpretation has been problematic. Armed with a useful purpose for such cases, such as the monitoring by facts of decisions, we are quite happy to allow factual falsification of value judgements, but without such a use it is difficult to assess the importance of the counter-examples.

We may now formalize the monitoring in some of the case studies considered earlier. In the case of the Manhattan Project it was originally decided to develop an atomic bomb if the Germans were working on it, on the principle that the Allies must be armed with any major weapon possessed by one of their enemies. But without the German threat, an atomic bomb project was seen as a colossal waste of resources which could be used to wage war. In other words the argument for development may be summarized:

> The allies ought to develop an atomic bomb if and only
> if one of their enemies is developing an atomic bomb.
> Germany is an enemy developing an atomic bomb.
> ―――――――――――――――――――――――――――――
> The allies ought to develop an atomic bomb. . . . 4

Monitoring was the gathering of intelligence about German developments, which eventually established that Germany did not have a bomb programme. This affects the decision because:

No enemy is developing an atomic bomb.

---

not-(The allies ought to develop an atomic bomb if and only if one of their enemies is developing an atomic bomb and The allies ought to develop an atomic bomb). ... 5

Here the factual premise entails the negation of the conjunction of two value judgements. Since the premise is accepted as true, one or both of the value judgements must be abandoned. It was argued in the discussion of the case study that the one which should have been rejected was the second, 'the allies ought to develop an atomic bomb', with the consequent abandonment of the Manhattan Project. In fact, of course, this was retained and the Project went ahead. This was only possible because the other value judgement was abandoned. It was maintained that the bomb might be worth developing even against an enemy not developing it, for use as an efficient military weapon or a way of imposing a war winning trauma on civilian populations.

The same logic applies to the breeder reactor decision. Here, the reason for building the reactors is to provide energy which can substitute for oil and monitoring is simply waiting to see how much energy the new technology can deliver. If, as in the discussion, the new reactors produce no energy something is wrong because:

Breeder reactors ought to be built only if they provide energy which can substitute for oil.
not-(Breeder reactors provide energy which can substitute for oil).

---

not-(Breeder reactors ought to be built). ... 6

Or, re-arranging:

not-(Breeder reactors provide energy which can substitute for oil).

---

not-(Breeder reactors ought to be built only if they provide energy which can substitute for oil and Breeder reactors ought to be built). ... 7

Here again, the factual premise entails the negation of a conjunction of value judgements. If the premise is accepted, therefore, at least one of the conjuncts must be abandoned. Either the decision to build breeder reactors must be recognized as wrong; formally, 'breeder reactors ought to be built' rejected, or else some other reason than producing energy as a substitute for oil must be found for building the reactors; formally, the rejection of the other conjunct. What makes the example so straightforward, of course, is that the first option seems much better than the second, unlike the situation in the Manhattan Project.

Having clarified the logic behind monitoring we may now turn to decisions where monitoring is not quite so straightforward as in the examples so far discussed. Here, it is always pretty clear cut when a decision is wrong, but for many decisions the question of whether they are erroneous is one of very hot debate. We need to ask what is the expected structure of such debates given the logical points so far discussed?

Suppose that a decision to do X has been taken and the associated project is underway. The question of whether the project should proceed, i.e. the question of the correctness of the decision to embark on it, is formally equivalent to whether the value judgement 'X is the best option' should be retained or rejected. Consider two parties, A who wants the project to continue and so holds 'X is the best option' and B who takes the opposite view, holding 'X is not the best option'. How can these two parties argue in a sensible way about the project's future? A and B could simply each assert their own value judgement, but this leads immediately to a head on clash of views which can make no progress. Something more than bald assertion is called for. We may suppose A and B to hold many value judgements in common, which we may call *background values* and denote $V_1 \ldots V_N$. Could A defend his position by showing that some set of these background value judgements, together with various factual claims, entail 'X is the best option'? In general this will be impossible. A is claiming that of all possible options, some of which may never have been explored in detail or never even made explicit, X is the best. His claim is therefore a very strong one which will not in general be deducible from any set of background values, which we can expect to be weak and pretty mundane just because they are shared. B, however, has an easier task because his claim is that some option is better than X. He can show this without having to demonstrate that the superior option is the best of all options. Because B's claim is much weaker than A's, what B can do to further his case is to search for some fact which, when coupled with some background values, say V_1, entails 'X is not the best option'. Formally, B should seek a factual sentence F for which there is some evidence and where

$$F \rightarrow \textbf{not-}(V_1 \text{ and } X \text{ is the best option}).$$

What B can then do is point out to A that F falsifies the conjunction of V_1 and 'X is the best option' so that A must amend one of these values, both of which he holds. If A accepts F and decides to retain V_1, he can do so only at the cost of rejecting 'X is the best option', in which case B has won the argument. A may, on the other hand, keep 'X is the best option' and reject V_1. This he cannot do, however, in an *ad hoc* way just to protect 'X is the best option' from criticism, but he must provide reasons for his rejection of V_1. In either case A is under stress because he must change his system of values. This stress may be eased, however, if A can deny F, the factual sentence giving rise to his worries. Again, this cannot be done in an *ad hoc* way just to save his cherished opinion about the project, but A must offer reasons for denying F. These reasons will, of course, be contended by B who will do whatever he can to convince A of F's truth.

It is now clear how the debate can progress and avoid collapse into a head on clash of opinions with no hope of resolution. Instead of a stubborn clash of values, the debate can proceed to consider the crucial question of whether F is true. This factual issue is only made relevant to the debate by the set of background values, but generally these values receive no explicit mention in the arguments. This is because they are shared and not expected to be changed in the debate, and shared because they are pretty trivial and mundane. Thus *all important issues over which an evaluative debate ranges are nearly always purely factual.* Although any debate about monitoring is an evaluative one – it is about what ought to be done, what is the best option – the main elements in the debate are always issues of fact. This is exemplified in the following Chapter which presents a case study of the debate between the US Environmental Protection Agency and the Ethyl Corporation about the correctness of the original decision to put lead into petrol and whether this should be prohibited in the future. Unlike earlier case studies this one has a Chapter to itself because it illustrates points made in this Chapter and also in Chapter 12 which follows it.

References

1 All the points discussed in this Chapter are dealt with more fully in Collingridge (1979) and (1980).
2 Ross (1939), p. 320.
3 Prichard (1949).
4 Geach (1958).

Bibliography

D. Collingridge (1979), 'The Fallibilist Theory of Value and Its Application to Decision Making', Ph.D. thesis, University of Aston.
(1980), 'The Autonomy of Values', *Journal of Value Inquiry, 14.*
P. Geach (1958), 'Imperatives and Deontic Logic', *Analysis, 18,* 49-56.
H. Prichard (1949), *Moral Obligation,* Oxford University Press.
W. Ross (1939), *The Foundation of Ethics,* Oxford University Press.

11. THE REMOVAL OF LEAD FROM PETROL

Case Study – US Environmental Protection Agency v. the Ethyl Corporation

This case study has a Chapter to itself because it illuminates the discussion of the two Chapters in between which it falls. It is a well documented and protracted disagreement between experts of the two bodies involved, the US Environmental Protection Agency and the Ethyl Corporation, concerning the health effects of lead from motor-vehicle exhausts on the American population. As such it can be seen as a disagreement about the monitoring of the original decision to permit lead additives in American petrol. As observed in Chapter 3, this decision was made in the belief that the health of the general population would be unaffected, so it is shown to have been in error if impairment of health from this source of lead is discovered.

Motor-vehicle exhausts have, for many years, been considered a serious nuisance and potential health hazard in the United States. Action was taken by Congress in 1970 to remedy these problems when a revision of the Clean Air Act was passed which placed severe limits on the gases causing the problems in vehicle exhaust. Under the amended Act, the Environmental Protection Agency (EPA) was given powers to control or prohibit the sale of any fuel which would endanger the public health or welfare or which would impair the performance of devices fitted to exhaust systems to control the level of noxious gases in accordance with the new limits. On 23rd February 1972, the EPA proposed a complex set of regulations which would require a phased reduction in the amount of lead added to petrol to 1.25 grms per gallon in 1977 and the general availability of a 91 octane (strictly a 91 Research Octane Number or RON) petrol free of lead by 1st July 1974 (37 Federal Register 3882-84 (1972)). The EPA's case was a double one. At the time no adequate pollution control device was available which could be attached to vehicles to reduce the level of noxious gases in their exhaust to below the new limits, but the most promising type of device employed a catalyst which would be unable to operate if lead was in the exhaust. For this reason EPA sought to ensure the availability of a lead-free fuel by 1st July 1974, from when new cars would have to meet the strict pollution requirements. The EPA also used their second authority to propose a reduction in the amount of lead added to petrol on the grounds that this constituted a public health hazard. They supported this claim with the document *Health Hazards of Lead*.

A major manufacturer of lead additives, the Ethyl Corporation, strongly resisted the EPA's attempt to gain approval for its proposed regulations. This case study traces some of the threads running through the long and complex debate between EPA and Ethyl. Although highly technical in many places, the debate offers a useful example to which to apply ideas on evaluation developed in the previous Chapter. It is very well documented and conducted by parties equipped with great resources, acumen and specialist knowledge. We might, therefore, expect to find in it examples of many of the points made earlier. Chief of these is the point that the debate, being about what course of action should be taken is a debate about values, and yet values play no explicit part whatever in the debate. The debate between Ethyl and EPA concerns facts and facts alone.

In any such debate, it is of the greatest importance to separate what the parties said from why they said it. In the case of Ethyl it is quite obvious what motivated them to attack the regulations proposed by EPA, for they faced ruin should they be put into effect. But this provides motivation only. The anguish of Ethyl lest they be ruined is not part of the critical debate, no more than EPA's angelical urge to trounce big business in the protection of the environment. Each side is motivated to argue its case as well as it can, but it is the argued case, and not the motivation which is important, at least for our present purposes.

Round 1

The debate between Ethyl Corporation and the EPA takes the form of a number of rounds, each side modifying its position in the light of the earlier round. The EPA open the first round with their publication of the proposed regulations to make lead-free fuels available and to gradually reduce the amount of lead added to normal petrol. Their first argument is that lead-free fuel must be available from 1st June 1974, when the new emission regulations come into force, so that catalytic emission control devices can be used:

> The administrator (of the EPA) has determined that emission products of lead additives will impair to a significant degree the performance of

emission control systems that include catalytic converters which motor vehicle manufacturers are developing to meet 1975-6 motor vehicle emission standards and are likely to be in general use if lead additives are controlled or prohibited for use in certain motor vehicle gasolines.[1]

EPA's argument may be captured formally in something like the following way:

V_1 The Federal regulations on vehicle emissions ought to be met.

F_1 The Federal regulations can only be met by catalytic systems by 1975.

F_2 Catalytic systems can only work if lead-free gasoline is available by 1975.

V_2 Lead-free gasoline ought to be available by 1975.

Ethyl replied to this argument by attacking F_1, the claim that only control systems employing catalytic converters would be working by 1975. It argued that it was unlikely that any workable control system would exist by 1975, so that implementation of the new standards would have to be postponed, involving a postponement of the regulations governing the availability of lead-free gasoline. Ethyl also argued that control systems which do not rely upon a catalyst might be found preferable to catalytic systems, and if these were used there would be no reason to insist that lead-free fuel be available to the motorist.[2]

Ethyl also attacked V_1, but not by simply denying it and substituting some other value statement of their free choosing. V_1 was criticized by appealing to facts. Ethyl argued that meeting the new emission standards would be very expensive and that using thermal reactors and not catalytic converters in vehicle exhaust systems would lead to slightly higher emissions than those allowed under the regulations, but would save a great deal of money. We might formalize their case as:

V_3 If X produces a marginal improvement in the environment at great cost then X ought not be done.

F_3 Meeting Federal emission regulations will produce a marginal improvement in the environment at great cost.

V_4 The Federal emission regulations ought not be met.

V_3 is a background value shared by both parties to the dispute and yet when coupled with F_3 it yields V_4 which contradicts V_1, assumed by the EPA to be a background value. If the EPA accept F_3, then they must reject at least one of V_3 or V_1. The conjunction $V_3.V_1$ is shown to be factually incorrect. If V_1 is rejected, then EPA's original argument collapses and Ethyl have won the argument. If the EPA wishes to counter this challenge, it should be obvious that this is best done by arguing against F_3. Thus the debate, although involving values, always centres on factual issues.

The second argument of the EPA was that lead from vehicle exhausts is a health hazard, so that the amount allowed in gasoline should be decreased. Very briefly, we may see this as:

V_5 If X eliminates a health hazard then X ought to be done.

F_4 Reducing the amount of lead in gasoline would eliminate a health hazard.

V_6 The amount of lead in gasoline ought to be reduced.

In support of F_4, EPA published the document *Health Hazards of Lead* (April 1972) largely based on a specially prepared report from the National Academy of Sciences, *Airborne Lead in Perspective* (1971). This concluded that:

F_5 Since lead has not been shown to have any biologically useful function in the body any increase in body burden of lead is accompanied by an increased risk of human health impairment.

F_6 In many cities air lead concentrations are slowly rising.

F_7 Human blood lead levels begin to rise appreciably with an exposure to airborne lead concentrations in excess of 2 micrograms per cubic metre.

F_8 Elevated lead intake for periods as short as three months produces an increase in blood lead levels.

F_9 Body burdens of lead increase with age, at least to forty years and probably thereafter.

F_{10} Although the ingestion of leaded paint is the predominant cause of lead poisoning in children, some children may show high blood lead levels from the ingestion of dust contaminated by fallout from airborne lead.

F_{11} Average blood lead levels tend to be higher among urban residents than among rural residents and higher among groups occupationally exposed to vehicle exhaust (e.g. policemen and garage workmen).

The Administrator of the EPA, therefore, recommended that:

> airborne lead levels exceeding 2 micrograms per cubic metre, averaged over a period of 3 months or longer, are associated with a sufficient risk of adverse physiologic effects to constitute endangerment of public health. Since airborne lead levels in many major urban areas currently range from 2 to somewhat over 5 micrograms per cubic metre, and since motor vehicles are the predominant source of airborne lead in such areas, attainment of a 2.0 microgram level will require a 60-65% reduction in lead emissions from motor vehicles.

Ethyl's reply to the case made out by the EPA on health grounds was very extensive, [3] but concentrated entirely on F_4, V_5 being accepted by both parties. Ethyl claimed that existing levels of lead in the air do not constitute a health risk, so that F_4 is false. In arguing this they proposed the following counters to the claims made by the EPA:

Against F_5 Years of experience with occupationally exposed groups show blood lead levels well in excess of those found in the normally exposed population to be perfectly safe.

Against F_6 The evidence indicates that air lead concentrations in many cities are falling. American blood leads are of the same order as those for many non-industrialized populations, indicating that lead from industrial sources makes only a small contribution to blood lead levels.

Against F_7 The data used in the calculation of this 2.0 microgram per cubic metre limit is seriously suspect, as is the statistical device used in the calculation. More reliable data (the so-called 7 City Study) shows no correlation between air lead levels and blood lead levels. In addition, the EPA assumed that about 30 per cent of lead inhaled is retained in the lung. The true figure is nearer 10 per cent.

Against F_8 This may be the case, but there is no evidence to indicate that the 'excess' blood lead levels resulting from exposure to airborne lead are a health hazard.

Against F_9 The data on body burdens shows that lead body burdens do not increase with age. Even if they do, this would merely reflect the very long time (about thirty years) needed for the body to come into equilibrium with environmental lead.

Against F_{10} There is no known correlation between lead levels in dust and earth and blood lead levels of children exposed to the dust and earth. There is no evidence whatever for EPA's hypothesis about dust being a significant contributor to the blood lead of children. The rate of lead fallout is so low that this can only be an insignificant source of lead.

Against F_{11} The very large 7 City Study reveals no correlation between air lead levels and blood lead levels.

In addition to offering these counters to the case made out by the EPA, Ethyl pointed to the economic and environmental cost of the EPA's proposal, arguing that these had been grossly underestimated by the EPA. The economic penalty would include, according to the Ethyl Corporation:

F_{12} 5 per cent more crude oil consumption due to using less efficient engines. This could put $1.4 billion on the balance of payment deficit.

F_{13} The cost to motorists will be about ¢4.7 per gallon because of higher gasoline costs and lower engine efficiency.

F_{14} Extra refinery investment to meet the need for low lead gasoline will amount to more than $4 billion. Small refineries will be unable to make the large investment necessary and will close.

If lead is not added to gasoline, the only way in which the intended octane number can be achieved is by adjusting the chemical mix of the product. This is done by altering the refining process (hence the refinery costs above) to give a gasoline containing a higher concentration of aromatic hydrocarbons. This would, according to Ethyl, constitute an environmental hazard since:

F_{15} Emission of polynuclear aromatic (PNA) hydrocarbons would increase, and many of these are suspected of causing human cancer.

F_{16} Emission of those hydrocarbons responsible for the formation of photo-chemical smog and eye-irritant chemicals would increase.

F_{17} EPA dismiss these problems as they expect these hazardous chemicals to be removed by the catalytic emission control systems to be fitted to new cars, but no working system yet exists. In addition, it is more difficult to work such a system when using fuel containing a high percentage of aromatic hydrocarbons.

We may formalize this part of Ethyl's case in the following way:

V_7 If X constitutes, on balance and at best, only a marginal improvement in human health at a great cost then X ought not be done.

F_{18} Reduction of lead levels in gasoline constitutes, on balance and at best, only a marginal improvement in human health.

F_{19} Reduction of lead levels in gasoline is very expensive.

V_8 Reduction of lead levels in gasoline ought not be done.

The force of the argument is as before. V_7 is a background value shared by both parties to the dispute. What Ethyl have done, therefore, is find facts, F_{18} and F_{19}, which, when coupled with some background value yield the evaluative consequence they want. If the factual claims are accepted by the EPA then they have no option but to revise their regulations calling for a reduction in the amount of lead added to gasoline. Returning to EPA's original argument ($V_5.F_4$ hence V_6), if they accept Ethyl's factual claims, then they must reject V_6 and hence at least one of V_5 and F_4 – presumably F_4. Formally, the conjunction $V_6.V_7$ is falsified by $F_4.F_{18}.F_{19}$. Ethyl have argued that F_{18} and F_{19} are true, so that their opponent can either accept $V_6.V_7$ and reject F_4 or else accept F_4 and reject either of V_6 or V_7. Since both V_6 and V_7 are background values, it seems that the rejection of F_4 is the preferred move, in which case EPA's original argument is bankrupted.

Ethyl's alternative to the EPA's regulations consists of ensuring that lead-free gasoline is available from when catalytic systems are known to work. New vehicles, from then on, would use lead-free fuel and so there would be a gradual elimination of leaded fuel. If some other emission control system is found preferable to the catalytic systems, then if it is still thought desirable to remove lead from the air, this can be done very cheaply by using existing fuels but fitting new cars with lead traps at present under development. Thus ended Round One of the contest.

Round Two

Following the criticisms of the Ethyl Corporation and others, the EPA revised their regulations. Previously, EPA's case for both parts of their proposed regulations, those governing the availability of lead-free gasoline and the phased reduction of lead concentrations in normal gasoline, were based on considerations of human health. In their revision the EPA now based the lead-free gasoline regulations solely on the need to have catalytic emission control devices and the regulations for the reduction of lead additives in gasoline on health considerations. The revised regulations were published on 10th January 1973 (Federal Register 1254-61), the health issues being published in a new document, *EPA's Position on the*

Health Effects of Airborne Lead, (9th March 1973). These two documents provided the main elements of Round Two of the debate.

The EPA dropped from their argument F_6 - F_9, the most important of which was F_7, the claim that blood lead levels rise on exposure to air containing more than 2 micrograms of lead per cubic metre. This was savagely attacked by Ethyl in the previous round and plays no more part in the debate. The EPA, however, reiterated $F5$, that lead has no known biological function so that any increase in body burdens increases the risk of human health impairment, F_{10}, the claim that airborne lead fallout in dust may be a significant route of lead exposure in children, and F_{11}, the claim that blood lead levels tend to be higher in urban residents than country residents, a claim which is now supported with evidence from the very large 7 City Survey. The following claims enter the debate for the first time:

F_{20} Many city dwellers have abnormally high blood lead levels.

F_{21} The susceptibility of children may be greater than adults so that children may be suffering subtle but unrecognized neurological impairment due to lead.

F_{22} Newborn babies in cities have higher blood lead levels than newborn babies in rural areas.

F_{23} Chromosomal damage due to lead is possible.

V_9 Presently recognized blood lead limits are too high to protect the public.

Upper acceptable limits for blood lead for the following groups are: expectant mothers, 30 milograms per 100 ml blood; newborn babies, 30; children, 40 (and perhaps less); adults, 40.

In support of F_{21} the EPA cited the work of David who compared the blood lead levels of children who were hyperactive with no known cause, with blood lead levels of a control group. He found that the hyperactive children had significantly higher blood lead levels than the controls. In support of V_9, the EPA referred to a recent study of umbilical cord lead levels.

Before considering the response made by the Ethyl Corporation, something must be said about the status of V_9. I have taken the statement as evaluative because it states that certain blood lead levels are the greatest which we ought to accept. EPA's argument for this involves appeal to a background value V_{10} in the following way:

V_{10} The upper acceptable limit for a toxin in the body is the lowest level at which the health of someone in the population is impaired.

F_{24} The lowest blood lead level at which the health of some expectant mother (newborn child, child, adult) is impaired is 30 (30, 40, 40) micrograms per 100 ml.

V_9 The upper acceptable limit for lead for expectant mothers (newborn children, children, adults) is 30 (30, 40, 40) micrograms per 100 ml in blood.

V_9 forms the most important part of the EPA's case and so it is not surprising to find that it comes under severe fire from the Ethyl Corporation. As before, the criticism centres not on the evaluative issues but on factual claims. Ethyl apparently accepts V_{10}, which may, therefore, be regarded as a background value, but it seeks to deny F_{24}, a move which would, of course, destroy EPA's argument for V_9. Ethyl denied that there was any medical evidence supporting F_{24}. The only evidence cited by the EPA concerned the study of umbilical cord blood levels, but this found no urban–rural gradient and concluded that there was no evidence to implicate airborne lead as a contributor to high cord blood lead levels. Ethyl accused the EPA of making an *ad hoc* move in so redefining upper acceptable blood lead levels without supporting evidence. They also pointed out that the Surgeon General regarded children with a blood lead level of less than 50 micrograms per 100 ml as safe provided there is no existence of continuing high lead exposure.

The Ethyl Corporation also countered the other parts of the case made out by the EPA:[4]

Against F_{20} The upper level for city dwellers' blood lead is around 40 micrograms per 100 ml which cannot be said to be 'abnormal'. Many higher values turn out to be the result of faulty analysis. For children, high blood leads are solely due to exposure to leaded paints.

Against F_{21} The major evidence for this is the work of David referred to earlier. Other investigations have failed to discover the same effect. David's results were due to the higher incidence of pica (the habitual eating of curious substances, often including lead paint) in hyperactive children.

Against F_{22} This claim is in direct contradiction to the paper cited by the EPA.

Against F_{23} The evidence for such chromosomal damage is extremely speculative.

Ethyl also expanded on some of the points in Round One of the debate. An extensive survey of lead poisoning in children failed to find a single case implicating dust as the source of lead. Ninety-eight per cent of cases were due to the eating of lead paint. Hence, EPA's claim that dust contaminated by fallout of airborne lead might be a significant source of lead exposure in children (F_{10}) is thrown in serious doubt. Ethyl also pointed to animal experiments indicating that lead might, after all, be an essential trace element, against EPA's claim F_5.

Finally, Ethyl argued that, on EPA's own admission, the removal of lead paint from delapidated, ageing housing would be considerably cheaper than the reduction of lead levels in gasoline. Such a programme of renovation would prevent the vast majority of existing cases of overt lead poisoning in children. We may formalize their case here in the following way:

V_{11} If X improves the health of the population less than Y and if X is more expensive than Y, then Y is preferable to X.

F_{25} Removing lead from gasoline will improve the health of the population less than removing lead paint from old buildings.

F_{26} Removing lead from gasoline will be more expensive than removing lead paint from old buildings.

V_{12} Removing lead from old buildings is preferable to removing lead from gasoline.

V_{11} is suggested as a background value, acceptable to Ethyl's opponent. If Ethyl can then show that F_{25} and F_{26} are true, then EPA must accept V_{12} and so give up V_6. To avoid this, EPA would, of course, attempt to counter one or both of these factual claims, so that, once again, the factual element of the debate is to the fore.

Round Three

EPA again revised their position on the health aspects of reducing the amount of lead added to gasoline in their final document, *EPA's Position on the Health Implications of Airborne Lead*, (28th November 1973) which was used to support their revised and final regulations of 6th December 1973. Unlike their earlier documents, this one was not open to public comment, and so Ethyl was not allowed a rejoinder. No doubt the EPA thought that the argument had gone on long enough. Despite this, however, the final document contained some major changes to the earlier one, several of them apparently the result of Ethyl's criticism.

1. It was admitted that there is no evidence suggesting that children are more susceptible to lead than adults. The special position of children is now based on their higher exposure to paint, dust and dirt.
2. The earlier recommendations for upper acceptable blood lead levels for expectant mothers and newborn children are dropped. Instead, limits for all are put at 40 micrograms per 100 ml of blood.
3. EPA argues that Ethyl's claim that there is no correlation between air lead concentrations and blood lead levels.

The EPA also expanded considerably on two of their earlier claims, that low levels of lead may cause subtle 'subclinical' neurological changes in children, and that contaminated dust is a major source of child exposure to the metal. They concluded that in each case the evidence was not conclusive, but taken together pointed to the correctness of their earlier claims. Quite new was a calculation of the likely increase in blood lead of a 'standard man' exposed to various air lead concentrations. These purported to show that to keep below the recommended blood lead of 40 micrograms per 100 ml, the ambient air lead concentration should be below 11.8 micrograms per cubic metre on optimistic assumptions and below 4.0 micrograms per cubic metre on pessimistic ones.

Round Four

Round Four takes the form of a court case. On the day that EPA promulgated their regulations, 6th December 1973, Ethyl petitioned the Court of Appeals to have them put aside. The case is of little value for our present purposes and a discussion of it would take us rather deeply into the fine points of law and precedence. Ethyl's submission to the Court, however, contains many criticisms of EPA's final health document, in particular on the sub-clinical effects of lead, on the correlation between air lead and blood lead levels and on the claim that dust containing lead is a hazard to children. Whilst of considerable interest in themselves, the arguments will add little illumination to our central concern – the nature of debates about monitoring.

On 28th January 1975 the Court of Appeals gave a majority finding in favour of Ethyl. The EPA then petitioned the Court for a rehearing which was granted and opened in May 1975. The case was a very interesting one, but to consider it in any detail would take us far from the point. Suffice it to say that the EPA won their case and that their final regulations were eventually approved and are now in force.

References

1 Federal Register (1972).
2 Catalytic exhaust systems are now fitted to all new American cars.
3 Ethyl Corporation (1972).
4 Ethyl Corporation (1973).

Bibliography

Environmental Protection Agency (1972), *Health Hazards of Lead.* (1973), *EPA's Position on the Health Effects of Airborne Lead.* (1973a), *EPA's Position on the Health Implications of Airborne Lead.*

Ethyl Corporation (1972), *Comments on EPA's Proposed Lead Regulations.* (1973), *Critique of 'EPA's Position on The Health Effects of Lead'.*

Federal Register (1972), 3882–84.

National Academy of Sciences (1971), *Airborne Lead in Perspective,* NAS.

12. THE ROLE OF EXPERTS IN DECISION MAKING

There is at the moment only the poorest understanding of the nature of expert advice and its role in the making of policy decisions in the kind of technical areas generally involved in the control of technology. Should one, for example, be concerned with the growing involvement of technical experts in determining the direction of major technological programmes, as in the Manhattan Project, or should this be welcomed as a long awaited improvement in the way that these decisions are made? To what extent can agreement between experts be expected? Is technical disagreement an inescapable and healthy feature of decision making, whose absence indicates inadequate scrutiny of some area of the decision; or is it to be seen as an aberration revealing the local collapse of scientific methodology? It is notorious that the opinion of an expert on some contentious question within his field, under debate by opposing interest groups, can often be predicted from knowledge of which group has his affiliation, but it is quite unclear what our reaction to this commonplace should be. Should the operation of bias, not to say corruption, be perceived here or should a more liberal view prevail? If the former, then what can be done to counter biased experts; if the latter, just how tolerant ought we be of such influences?

The purpose of this Chapter is to derive some answers for these and similar questions, drawing on the theoretical discussion of Chapter 10 and the case study of the previous Chapter. It is best to begin by considering a model, which I shall call *Model 1*, which underlies much thinking about the role of expert opinion. I want to argue that Model 1 is very misleading and in urgent need of replacement.

Model 1 is centred on a very simple view of science. It sees science as involving the collection of *data* from experience and observation, which is used as a base for the construction of theories, so that for any set of data there is a single *interpretation* or theoretical explanation, of the data which the canons of scientific method show to be superior to all others.

Since all experts are armed with the same appreciation of scientific method, they can be expected to agree on a single interpretation of any set of data which is offered to them, all rival interpretations being rejected by them. On this view, disagreement between experts is not to be welcomed. At best it marks an inadvertent failure to apply the methods of the science in question correctly, or to consider all the relevant data; at worst, it indicates bias or corruption in one or other of the debating experts.

This feature of Model 1 makes it particularly attractive to decision makers who have to rely on expert opinion but who have, themselves, no insight into the workings of scientific method. Such a decision maker obviously desires a consensus from his experts, for disagreement merely adds to his problems as the options which are open expand instead of diminish. Moreover, such a person will tend to view the enterprise of science in what we shall see to be a distortedly simple way. For him, science is the provider of knowledge, so that the persistence of disagreement is a blotch amounting to the local collapse of science. For people trained in isolation from science the vaccilation of scientists as they carefully enumerate the various hypotheses which might be upheld in the light of the evidence available, often seems an unnecessary irritation, especially where it complicates already complex decisions.

Deciding what to do requires facts and values. If two parties disagree about what is to be done, this may be because they accept different values or because they have different beliefs about the facts which are relevant to the decision. According to Model 1 if both parties are advised by experts who have all the data open to them, they must both accept the same interpretation of the data so that a continuation of the disagreement about what to do can only come from the acceptance of different value judgements. In the case study of the previous Chapter, for example, disagreement between the US Environmental Protection Agency (EPA) and the Ethyl Corporation about whether lead should be removed from petrol cannot arise from disagreements over the facts which are relevant, since both sides have access to experts in all the required fields and these experts will come to the same opinion on each issue before them. The disagreement, instead, arises from the different values adopted by the two parties. Both agree that lead from this source is a danger to the health of children, but the Ethyl Corporation favours retaining lead in petrol because it considers its own profitable existence to be more valuable than the health of children. The EPA, on the other hand, operate with value judgements designed to protect the environment, so in this case they favour the removal of the health hazard from exhaust fumes over the continuation of the Ethyl Corporation. Figure 12.1 summarizes the view which Model 1 gives of the disagreement.

Figure 12.1
Model 1 Applied to the Debate Between the EPA and Ethyl Corporation

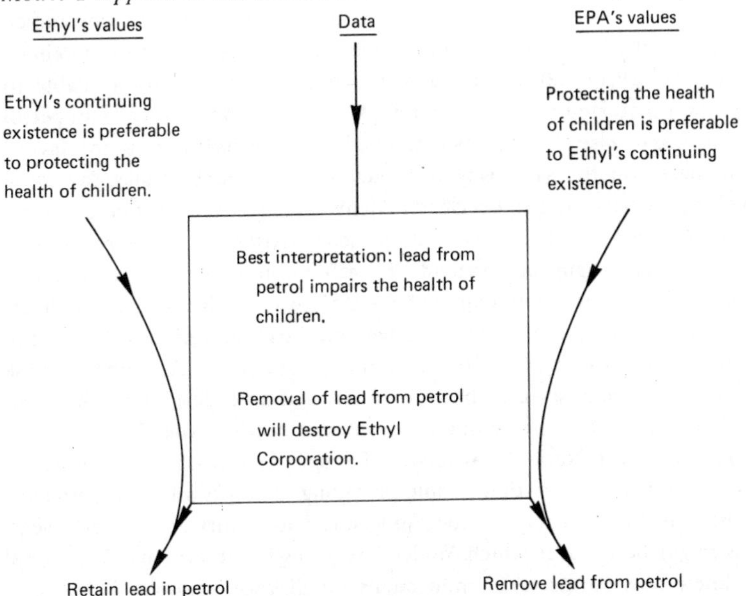

I now wish to bring the following criticisms of Model 1.

(a) Where the model is put against a real case, as above, it is immediately apparent that the mismatch is almost total. In the debate between Ethyl and EPA the values supposedly at the root of their disagreement were never discussed, or even mentioned, nor any values resembling them in the weakest respect. Values simply did not figure explicitly in any of the arguments between the parties. Nor, of course, did the rivals agree on the interpretation of the data concerning the effects of lead from vehicle exhausts on the health of children. EPA claimed that the best interpretation was that lead from this source impairs the health of children, but the Ethyl Corporation held that the best interpretation of the selfsame body of data was that there is no effect on the health of children. The experts employed by each party failed to reach agreement in all important cases, totally counter to the view of expert opinion embodied in Model 1. Real disagreements about what should be done are all like this; conflict concerns not values but what interpretation of the data is to be favoured. On Model 1 the case of the EPA v. Ethyl should be highly atypical, but it is not. Most important technological decisions are favoured by one interest group and heavily opposed by another, and the debate between them, when it happens, has the same form as the debate between the Ethyl Corporation and the EPA.

(b) Model 1 can only account for the failure to reach agreement about the best interpretation of the data in such cases as the one considered above, by attributing bias to one, or both of the parties to the debate. Either the EPA's experts are biased and interpret the data available to them in a distorted way to favour EPA's contention that lead from petrol is harmful, or else the experts employed by the Ethyl Corporation favour their master in the same way. Although extreme cases of bias, say those involving bribery, might be clearly identified, it is very difficult to establish any more sensitive test for its presence. After all, bias is not simply of pecuniary origin but extends to affiliation to colleagues past and present, disaffection with experts belonging to some rival group, the desire for publicity and fame in the world beyond the laboratory, the acquisition of habits of research in one's early training, and so on. Any one of these and their like may serve to bias the opinion of an expert, and so how can we then hope to find a true opinion from an objective expert?

The American National Academy of Sciences study of science in public policy making suggests that people providing scientific advice should have qualities of 'intelligence, wisdom, judgement, humanity and perspective'.[1] This might be the test which Model 1 requires for absence of bias, if only we knew how to assess an individual's intelligence, wisdom, judgement, humanity and perspective. Dissension over who has what qualities is likely to be even more acrimonious and irresoluble than the original technical disagreement, which at least does not concern reputation and personal attributes. No doubt the only people to judge of these angelic properties are those who themselves possess intelligence, wisdom, judgement, humanity and perspective, although they might be somewhat slow in reporting their findings as there seems an infinite regress for them to work through. The body publishing this list of attributes also demands a 'bias statement' from scientists who provide information to the American government. This document lists all previous major research funding, occupations, investments, public stands on political issues, and anything which 'might appear to other reasonable individuals as compromising of your independence of judgement'.[2] How such lists are to be marked on the scale of bias is, however, a complete mystery. Their value to the decision maker is also rather suspect. Already facing a mountain of information, his desk is further loaded with dissenting reports swelled by hundreds of 'bias statements'. Just what is he supposed to do with this indigestible mass?

Because there are no straightforward tests for the presence of bias it is easy to see it everywhere. In Boffey's study of the advice given to the American government by members of the National Academy of Sciences, an individual's background is scratched over until some way is seen of connecting his private interest and the advice he has given, this being seized upon as evidence of bias.[3] The *reductio ad absurdum* is neatly performed by Polsby in his review of Boffey's book.[4] Polsby simply points out that Boffey was employed by Ralph Nader to write his book and that Nader sees his task as exposing 'establishment-type establishments' so that anyone working for him is bound to be biased in their investigation of such an establishment. All that is needed now of course, is someone to investigate Polsby's background, and so on, and so on.

The problem is that Model 1 provides a psychological, and so a covert and unexplorable view of bias. Bias is in the mind of the scientist. The data points to a uniquely favoured interpretation, and if this is not accepted by a scientist who is competent and acquainted with the data, the only cause can be bias. Hammond and Adelman have rightly objected to this concept because it makes expert judgement a total mystery. What are the mechanisms by which an unbiased mind can be achieved, how can it be eliminated if it is present and what public tests are there for it? The mistake is to personalize the problem. In the words of Hammond and Adelman:

> It is precisely because scientists have learned that it is not only fruitless, but harmful, to focus on persons and their motives that they have learned to ignore them in their work as scientists. When scientists look for the truth and the truth appears to be in doubt, neither scientific work nor the scientific ethic requires the investigation of the characteristics of the person working on the problem; instead, they require the analysis of the method by which the results are produced.[5]

(c) The third criticism of Model 1 concerns the relationship it supposes to hold between experts' advice on technical matters and their role in policy making. The expert, on Model 1, is a mere provider of technical information and has no special role in the making of decisions. A scientist employed by the Ethyl Corporation in their fight against the EPA's plans to remove lead from petrol serves purely to provide his employer with data and to interpret data from other sources. That his professional opinions are relevant to the survival of the Ethyl Corporation and to the health of millions of children are of no concern to the expert. He may not share the values of either the Ethyl Corporation or the EPA, so that his expert opinions may point to no course of action at all as far as he is concerned. If he does share the values of one of the parties to the debate, then this is not because he is an expert, it is part of his non-professional existence. Thus the unbiased expert is, as professional expert, neutral between interested parties arguing what is to be done and disinterested in what decisions are made on the basis of his opinions. This, if you like, is Model 1's *black box* view of expert opinion. It is a widely shared view as witnessed by the following remarks by Phillip Handler, then President of the US National Academy of Sciences:

> Science can contribute much to enhancing agricultural production, but American policy with respect to food aid is not intrinsically a scientific question. Similarly, science can study whether energy independence is technically feasible or whether Soviet underground nuclear tests can be detected, but ... (scientists) must then let regular policy-makers decide whether to try for energy independence or just what arms control proposals to put to the Russians.[6]

Is the division into the two realms of science and policy really as clear cut as Handler suggests? The Manhattan Project has a lesson here. Scientists are drawn into questions of policy just because the making of policy decisions requires more than the digestion of delivered facts, it calls for a dialogue between expert and decision maker. The latter may need to explore options whose existence is known only to a handful of experts, while the experts need to be advised by the policy maker about what he thinks are the critical unknowns in the present state of policy. In the words of Boulding:

> Every decision involves the selection among an agenda of alternative images of the future, a selection that is guided by some system of values. The values are traditionally supposed to be the cherished preserve of the political decision-maker, but the agenda, which involves fact or at least a projection into the future of what are presumably factual systems, should be very much in the domain of science ... (but) if the decision-maker simply does not know what the results of alternative actions will be, it is difficult to evaluate unknown results. The decision-maker wants to know what are the choices from which he must choose.[7]

The clear cut division seen by Handler is, therefore, like so many similarly perceived divisions, an oversimplification.

Having considered the shortcomings of Model 1, I now wish to propose a superior view, that of Model 2. On this model the data of science, that is its collected observations and experimental results, is still recognized, but it is admitted that any set of data will be able to bear a number of conflicting interpretations. In a debate about what should be done both sides share a common fund of what may be called background values, and these are the ones to which they will appeal in making out their case. What each side tries to do is to find an interpretation of the agreed data which is scientifically respectable and which leads, when coupled with background values, to the desired course of action. The debate between the parties over what should be done will, therefore, centre on which side offers the best interpretation of the data, and will not explicitly concern evaluative issues at all. This, of course, is the same sort of debate as goes on in science all the time, because one of the functions of science is the elaboration, testing and selection of rival interpretations of data. The debate about what to do therefore becomes like any similar debate in science, to be won or lost by the rules of scientific method.

An example is given in Figure 12.2 which outlines the analysis offered by Model 2 of the debate about removing lead from petrol. The background values are such mundane things as 'lead should be removed from petrol only if it impairs the health of children' and 'if lead in petrol impairs the health of children then it should be removed'. Both of these are accepted by the Ethyl Corporation and the EPA who each then seek an interpretation of the scientific data which will couple with these background values to generate the option which is desired. The EPA tries to show that the best interpretation is that lead is hazardous, so that it ought to be removed, and its rival defends the interpretation that lead is harmless, so that it ought not to be removed. The two interpretations fight it out according to the general rules of scientific method.

The model makes a clear distinction between this *public argument* and the *motivation* behind it. The motivation for Ethyl's argument is obviously the Corporation's continued existence, but this is not an element in the public case which it makes out. Ethyl cannot argue that lead should remain in petrol so that it can continue to exist, because no other party to the debate has the slightest interest in the Corporation's future. No more can the EPA argue that lead should be removed as a way of showing how effectively the EPA protects the environment against big business, because nobody else in the debate cares a fig for EPA's image and status. If argument is to be more than the wasting of breath, it must appeal to values common to all parties, i.e. to background values. This, of course, is what happens in the public debate between Ethyl and the EPA, but there would be no reason for the debate at all if one side was not motivated in some way, so that it had no preferred option. Rival motivation ensures that the debate is carried out energetically and that the best case is made for each interpretation and that all suggested interpretations receive intense scrutiny. This, of course, is essential if the final decision is to be a good one, one based on a thoroughgoing scientific debate about the best interpretation of the data on the health effects of lead from vehicle exhausts on children.

Figure 12.2
Model 2 Applied to the Debate Between the EPA and Ethyl Corporation

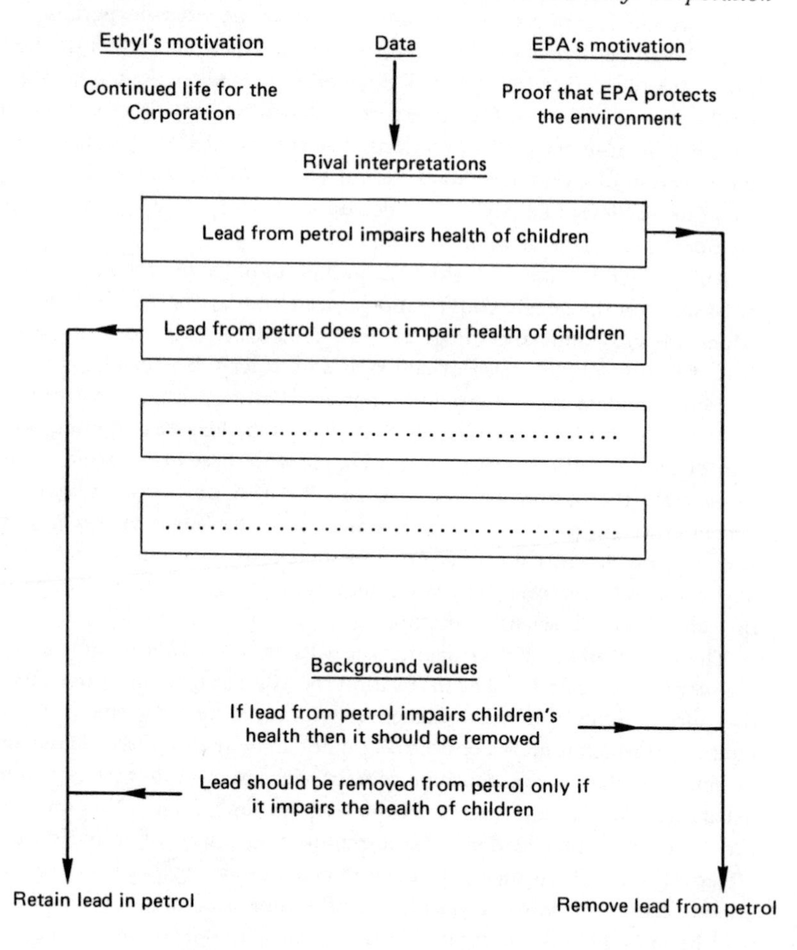

The new model gives a much more satisfactory account of expert advice than that offered earlier. First of all it gives a much more realistic account of disagreements about what action to take. As observed in the previous two Chapters, debates of this kind have very little concern for evaluative issues, the key elements in the debate being factual and concerning the best interpretation of scientific data. This accords exactly with Model 2. In addition, Model 2 also explains why disagreement between experts is so common in the making of complex decisions. This does not reveal how common bias is within the community of experts, which is the only explanation open to a supporter of Model 1. On Model 2 data can be interpreted in a number of ways and different experts will favour different interpretations which they can then fight over. Disagreement and debate is not at all shocking, it is a sign of a healthy and exploring science, searching for the best way of seeing some set of data. What Model 2 shows is that it is urgently necessary to develop a theory of decision making *which can accommodate the fact that experts can be expected to disagree*. More of this later.

The problem with Model 1's account of bias was that the existence of bias becomes a hidden, psychological feature of the individual expert. On Model 2 it is possible for debates about what to do to be biased, but in a publicly discoverable way. For this model such a debate is centrally an argument between proponents of rival interpretations of some body of data. Unfairness can arise here from a number of sources; from massively unequal funding of the parties so that the scale of expert advice available to each is very different, from the domination of research by a few experts who place work in the field along rigid lines so that rival interpretations are never explored, and from administrative rules which impose secrecy or which place certain key factual claims in the debate beyond argument. The first source of bias is very common where the debate occurs within the context of a public inquiry. The proposer is often a large private or public body with its own research staff and able to buy expert opinion from outside as well, while the objectors may be individuals or small protest groups hastily thrown together and having no research skills or money to buy opinions.[8] Such inquiries also often exemplify bias from the third source, as they are often held under rules which make it impossible for some of the data to be discussed because it is secret, or which make it impossible to question certain claims. A notorious example of this last is road building inquiries where the need for the proposed road cannot be discussed, merely whether it is the best route.[9]

The second mentioned origin of bias deserves a little more attention than the others because it is more subtle, so we may consider Kehoe's early work on the toxicity of lead as an example. Kehoe did much of the fundamental work on the behaviour of lead in the human body, and was provided with facilities, mostly funded by the lead industry, unequalled by other laboratories. Not surprisingly much of Kehoe's experimental and analytical work was not repeatable at the time, so that his work tended to fix the direction in which the whole field was going. Kehoe was an industrial hygienist, concerned with lead exposure in industry. He thus worked only on men, and men of working age who were fit. Whole areas of investigation were, therefore, put on one side, not just by Kehoe but by researchers generally; for example, the effects of lead on women, children, the foetus and on people with various diseases.

This is clearly seen in the famous controversy about Kehoe's 80 μg/100ml blood lead threshold limit, which has already been briefly mentioned in the previous Chapter. Kehoe stated that in his wealth of experience no person had shown signs of clinical lead poisoning from an exposure which had produced a blood lead concentration of less than 80 μg/100 ml, so that this could be viewed as a threshold limit, below which safety is guaranteed. Kehoe's experience, however, was with healthy, active men under medical supervision, as all lead workers are, and moreover, ones who remain in a population screened for unusual sensitivity to lead. If a new worker shows signs of poisoning or if his blood lead level becomes unexpectedly high, he is moved from the job to one where he will receive a lower exposure to the metal. In this way the remaining workers are *hypo*-sensitive to lead. Nothing which could be said about this group could be said about the general population, which includes males who would be filtered out if they worked with lead, sick men, old men, women and children. But such was Kehoe's standing in the scientific community and so dominant was his work that his threshold limit was taken as applying to the whole population, which was then taken to be totally safe from toxic effects of lead.[10]

Bias of the kind identified by Model 2 may be subtle, or it may be quite explicit, but in either case it does not require investigation of the psychology of individual experts. It is a public, overt question whether bias exists, and when it is found it may be remedied by public methods such as altering the rules of public inquiries, allocating funds more evenly and ensuring that no Kehoe dominates the research area which is relevant to the debate.

Another problem for Model 1 was its simple-minded division between advice and policy which we saw to be untenable. Here again, Model 2 performs better. On Model 2 the advisor can be seen as much more of an advocate, actively engaged in the policy debate. The sort of values which figure in such a debate are generally so mundane and ordinary that only the wildest eccentric would want to question them. These values are shared by the parties to the debate, but can be expected to be also held by any of the experts they consult. The expert's task is to offer and defend interpretations of the data which, when coupled with these values, lead to the policy option favoured by his master. The interpretation is the key item, it is this which is disputed, not the value judgements. The expert can no longer pretend to be above the conflict, saying 'I just give the facts; what values are used to turn them into decisions is not my business'. He will generally accept the background values shared by the debate's parties so that if he is convinced that a particular interpretation which favours one side is the best, then he will himself regard the policy option supported by that side to be the best. The expert is, therefore, inside the debate and cannot help but be there.

There is nothing wrong in seeing an expert as an advocate. His task is the perfectly proper one of discovering and defending interpretations of the data which are in accordance with scientific method, to take extreme examples, he must not destroy recalcitrant data or invent data supporting his position, and which also serve the interest of his master by yielding the policy conclusions he wants when conjoined with background values. This would be pure ritual if it were not for a rival group of experts seeking to serve a master who favours an opposing policy. The battle is then fought by trying to discover which of the expert groups provides the best interpretation, the rival motivation of the paymasters ensuring that the debate about this is thorough and deep.

Having seen the superiority of Model 2 over Model 1 we may now take up a point from the earlier discussion; if experts are likely to disagree how can decisions be made on the basis of expert opinion? A very common response is that the decision should be postponed until a consensus is reached. Enough has been said already to reveal the shallowness of this approach. It is perfectly in order where delay has no cost, but these are rare cases indeed. Generally delay is expensive and its expense must be traded against the chance of making the wrong decision before the experts have reached a consensus. The discussion of removing lead from petrol in Chapter 2 will no doubt be recalled here.

A more sophisticated suggestion is that a 'science court' should hear the arguments of the various experts and then decide between them. A detailed suggestion along these lines was made to the US Congress by Kantrowitz, who urged them to 'appoint a science advocate for (each) side of the story', the procedure being 'modelled on the judicial procedure for proceeding in the presence of scientific controversy'. Final judgement would be from a group of scientific judges who would argue out the case between them.[11] The suggestion has been heavily criticized, but it must be recognized that it tries to answer a genuine problem. Some way is needed of handling expert disagreement, because this can be expected to be the rule rather than the exception. Where the idea fails is in being welded to the belief that a decision requires hard information. If this cannot be generated by a consensus among the relevant experts, then it must be provided by surviving the critical scrutiny of a science court.

What is needed, on the contrary, is some way of making the decision in the absence of consensus so that it can be changed if later research leads to an agreement between the experts that shows the original decision to have been wrong. In the absence of consensus it is essential to preserve the decision maker's ability to detect error in his decision and his ability to correct it. The scientific debate between the experts should not, therefore, halt when a decision is made, it should continue because a consensus may be reached which shows the original decision to have been wrong, so that it must be corrected. This, of course, is the monitoring of the original decision which has had to be taken under ignorance, given the split among the experts. Options which are highly flexible, insensitive to error, and easy to correct should, therefore, be favoured.

In the case study of the previous Chapter there is no consensus about the health effects of lead from vehicle exhausts, so that any decision must be taken under ignorance. It therefore calls for monitoring, i.e. for the continuation of the debate about health effects. If, for example, lead in petrol is retained then a future consensus that lead from this source damages the health of children would show the decision to have been wrong, and it should be corrected, by deciding to remove lead from petrol. If the first decision is to remove the metal from petrol and a consensus later forms that it is harmless, the decision may be corrected by allowing the addition of lead to petrol once more. Ensuring that it is possible to revise the decisions in this way may, of course, be expensive, and the cost of continuing the debate about health effects will add to this cost, but good decision making is not free. In this way Model 2 can handle cases where no single interpretation of the data emerges as the best.

References

1 For a critical summary see Skolnikoff and Brooks (1975).
2 See Boffey (1975).
3 Boffey (1975).
4 Polsby (1975).
5 Hammond and Adelman (1976), p. 391.
6 Handler (1975).
7 Boulding (1975), p. 423. See also Toulmin (1972).
8 A spectacular example is the recent Windscale inquiry. Discussions of fairness are depressingly few in the literature, see Council for Science and Society (1976) and (1979) and Rickson (1976).
9 See Law Report (1980).
10 Kehoe *et al.* (1933).
11 Kantrowitz (1971).

Bibliography

P. Boffey (1975), *The Brain Bank of America*, McGraw-Hill.
K. Boulding (1975), 'Truth or Power', *Science*, *190*, 423.

Council for Science and Society (1976), *Superstar Technologies*, Barry Rose.
(1979), *The Big Public Inquiry*, Barry Rose.

K. Hammond and L. Adelman (1976), 'Science, Values and Human Judgement', *Science*, *194*, 389-96.

P. Handler (1975), Interview, *Wall St. Journal*, 3rd April, 12.

S. Kantrowitz (1971), *Hearings before the House Committee on Rules and Administration*.

R. Kehoe *et al.* (1933), 'On the Normal Absorption and Excretion of Lead', *Journal of Industrial Hygiene*, *15*, 273-88.

Law Report (1980), *Bushell and Another* v. *Secretary of State for the Environment*, House of Lords, *The Times*, 11th February.

N. Polsby (1975), Review of P. Boffey, *Science*, *190*, 665.

R. Rickson (1976), 'Knowledge Management in Industrial Society and Environmental Quality', *Human Organization*, *35*, 239-51.

E. Skolnikoff and H. Brooks (1975), 'Science Advice in the White House – The Continuation of a Debate', *Science*, *187*, 35-41.

S. Toulmin (1975), in *Civilisation and Science: In Conflict or Collaboration*?, Ciba Foundation Symposium, Elsevier.

INDEX